Research Design and Methodology

Research Design and Methodology

Edited by
Lane Cobb

Research Design and Methodology
Edited by Lane Cobb
ISBN: 978-1-63549-250-7 (Hardback)

Published by Larsen and Keller Education,
5 Penn Plaza,
19th Floor,
New York, NY 10001, USA

Cataloging-in-Publication Data

Research design and methodology / edited by Lane Cobb.
p. cm.
Includes bibliographical references and index.
ISBN 978-1-63549-250-7
1. Research. 2. Research--Methodology. I. Cobb, Lane.
Q180.A1 R47 2017
507.2--dc23

Printed and bound in the United States of America.

For more information regarding Larsen and Keller Education and its products, please visit the publisher's website www.larsen-keller.com

Table of Contents

Preface

Research design refers to the analysis of different study types like experimental, correlational, review, descriptive, meta-analytic, hypothesis, case study, data collection, independent and dependent variables, statistical analysis, etc. It is designed to find answers to research questions. Through this book, we have attempted to provide students a detailed explanation of the various fundamental concepts of research design and methodology. It picks up individual concepts and explains their need and contribution in the context of the growth of the subject. The topics introduced in the text cover the basic principles of the subject. The textbook aims to serve as a resource guide for students and facilitate the study of the discipline.

A detailed account of the significant topics covered in this book is provided below:

Chapter 1- Research design is the plan that is used in a particular field. It helps in understanding the study type, which is usually either descriptive, experimental, review or meta-analytic. The chapter on research design offers an insightful focus, keeping in mind the complex subject matter.

Chapter 2- Case study is the study that concentrates on a particular person, group or situation. They can be written in the form of journals or presented in professional conferences. Some other types of research design included in this chapter are naturalistic observation, human subject research, causal research, correlation and dependence and scientific method. Research design is best understood in confluence with the major topics listed in the following chapter.

Chapter 3- Social research is a research that is conducted by students or social scientists. The topics explained in the section are quantitative research, multimethodology, academic publishing, peer review, clinical peer review etc. The aspects elucidated in this chapter are of vital importance, and provide a better understanding of social and academic research.

Chapter 4- The methods used in research are ethnography, grounded theory, field research and meta-analysis. Ethnography is the study of people and of culture. This study has a holistic approach and also studies popular subjects such as sociology, history and communication studies. The section discusses the methods of research in a critical manner providing key analysis to the subject matter.

Chapter 5- The important aspects of research are hypothesis, theory, law, conceptual model and generalization. Hypothesis is an explanation that is suggested for a particular phenomenon whereas theory is an abstraction which is rational in its approach and tries to explain thought. The chapter serves as a source to understand the major classifications related to the fundamentals of research.

Chapter 6- Observational study is the study of a sample to a population. In this study the subject is not under the control of the researcher mainly because of ethical concerns. The other essential aspects that have been explained are experiments, field experiments, literature review and systematic review. The major categories of research design are dealt with great detail in the chapter.

Chapter 7- The ethics that are involved in researches are known as research ethics. It includes topics such as ethical research in social science, clinical research ethics, committee on publication ethics and privacy for research participants. This section has been carefully written to provide an easy understanding of the varied facets of research ethics.

Chapter 8- Methodology is an organized analysis of the methods that are applied to a particular field of study. It mainly consists of concepts such as paradigm, theoretical model and quantitative techniques. The other topics explained are art methodology and rationalism. In order to completely understand methodology, it is necessary to understand the themes elucidated in the following section.

Chapter 9- The key concepts in methodology are research question, thesis statement and paradigm. The research question is the question that is asked in the research paper, which the research paper answers as its conclusion. In order to completely understand methodology, it is necessary to understand the processes related to it. The following chapter elucidates the main concepts in methodology.

Chapter 10- Philosophical methodology is the method of doing philosophy. Some of the methods that philosophers follow are methodic doubt, argument and dialectic. The themes discussed are phenomenology, quietism, methodism, experimental philosophy and intercultural philosophy. These topics are crucial for a complete understanding of the subject of research design and methodology.

I would like to make a special mention of my publisher who considered me worthy of this opportunity and also supported me throughout the process. I would also like to thank the editing team at the back-end who extended their help whenever required.

Editor

1

Introduction to Research Design

Research design is the plan that is used in a particular field. It helps in understanding the study type, which is usually either descriptive, experimental, review or meta-analytic. The chapter on research design offers an insightful focus, keeping in mind the complex subject matter.

A research design is the plan of a research study. The design of a study defines the study type (descriptive, correlational, semi-experimental, experimental, review, meta-analytic) and sub-type (e.g., descriptive-longitudinal case study), research question, hypotheses, independent and dependent variables, experimental design, and, if applicable, data collection methods and a statistical analysis plan. Research design is the framework that has been created to seek answers to research questions.

Design Types and Sub-types

There are many ways to classify research designs, but sometimes the distinction is artificial and other times different designs are combined. Nonetheless, the list below offers a number of useful distinctions between possible research designs. A research design is an arrangement of conditions or collections.

- Descriptive (e.g., case-study, naturalistic observation, survey)
- Correlational (e.g., case-control study, observational study)
- Semi-experimental (e.g., field experiment, quasi-experiment)
- Experimental (experiment with random assignment)
- Review (literature review, systematic review)
- Meta-analytic (meta-analysis)

Sometimes a distinction is made between "fixed" and "flexible" designs. In some cases, these types coincide with quantitative and qualitative research designs respectively, though this need not be the case. In fixed designs, the design of the study is fixed before the main stage of data collection takes place. Fixed designs are normally theory-driven; otherwise, it is impossible to know in advance which variables need to be controlled and measured. Often, these variables are measured quantitatively. Flexible designs allow for more freedom during the data collection process. One reason for using a flexible research design can be that the variable of interest is not quantitatively measurable, such as culture. In other cases, theory might not be available before one starts the research.

Grouping

The choice of how to group participants depends on the research hypothesis and on how the participants are sampled. In a typical experimental study, there will be at least one "experimental" condition (e.g., "treatment") and one "control" condition ("no treatment"), but the appropriate method of grouping may depend on factors such as the duration of measurement phase and participant characteristics:

- Cohort study
- Cross-sectional study
- Cross-sequential study
- Longitudinal study

Confirmatory Versus Exploratory Research

Confirmatory research tests *a priori* hypotheses — outcome predictions that are made before the measurement phase begins. Such *a priori* hypotheses are usually derived from a theory or the results of previous studies. The advantage of confirmatory research is that the result is more meaningful, in the sense that it is much harder to claim that a certain result is generalizable beyond the data set. The reason for this is that in confirmatory research, one ideally strives to reduce the probability of falsely reporting a coincidental result as meaningful. This probability is known as α-level or the probability of a type I error.

Exploratory research on the other hand seeks to generate *a posteriori* hypotheses by examining a data-set and looking for potential relations between variables. It is also possible to have an idea about a relation between variables but to lack knowledge of the direction and strength of the relation. If the researcher does not have any specific hypotheses beforehand, the study is exploratory with respect to the variables in question (although it might be confirmatory for others). The advantage of exploratory research is that it is easier to make new discoveries due to the less stringent methodological restrictions. Here, the researcher does not want to miss a potentially interesting relation and therefore aims to minimize the probability of rejecting a *real* effect or relation; this probability is sometimes referred to as β and the associated error is of type II. In other words, if the researcher simply wants to see whether some measured variables could be related, he would want to increase the chances of finding a significant result by lowering the threshold of what is deemed to be *significant*.

Sometimes, a researcher may conduct exploratory research but report it as if it had been confirmatory ('Hypothesizing After the Results are Known', HARKing—; this is a questionable research practice bordering on fraud.

State Problems Versus Process Problems

A distinction can be made between state problems and process problems. State problems aim to answer what the state of a phenomenon is at a given time, while process problems deal with the change of phenomena over time. Examples of state problems are the level of mathematical skills of sixteen-year-old children or the level, computer skills of the elderly, the depression level of a person, etc. Examples of process problems are the development of mathematical skills from puberty to adulthood, the change in computer skills when people get older and how depression symptoms change during therapy.

State problems are easier to measure than process problems. State problems just require one measurement of the phenomena of interest, while process problems always require multiple measurements. Research designs such as repeated measurements and longitudinal study are needed to address process problems.

Examples of Fixed Designs

Experimental Research Designs

In an experimental design, the researcher actively tries to change the situation, circumstances, or experience of participants (manipulation), which may lead to a change in behavior or outcomes for the participants of the study. The researcher randomly assigns participants to different conditions, measures the variables of interest and tries to control for confounding variables. Therefore, experiments are often highly fixed even before the data collection starts.

In a good experimental design, a few things are of great importance. First of all, it is necessary to think of the best way to operationalize the variables that will be measured, as well as which statistical methods would be most appropriate to answer the research question. Thus, the researcher should consider what the expectations of the study are as well as how to analyse any potential results. Finally, in an experimental design the researcher must think of the practical limitations including the availability of participants as well as how representative the participants are to the target population. It is important to consider each of these factors before beginning the experiment. Additionally, many researchers employ power analysis before they conduct an experiment, in order to determine how large the sample must be to find an effect of a given size with a given design at the desired probability of making a Type I or Type II error.

Non-experimental Research Designs

Non-experimental research designs do not involve a manipulation of the situation, circumstances or experience of the participants. Non-experimental research designs can be broadly classified into three categories. First, in relational designs, a range of variables are measured. These designs are also called correlation studies, because correlation data are most often used in analysis. Since correlation does not imply causation, such studies simply identify co-movements of variables. Correlational designs are helpful in identifying the relation of one variable to another, and seeing the frequency of co-occurrence in two natural groups. The second type is comparative research. These designs compare two or more groups on one or more variable, such as the effect of gender on grades. The third type of non-experimental research is a longitudinal design. A longitudinal design examines variables such as performance exhibited by a group or groups over time.

Examples of Flexible Research Designs

Case Study

Famous case studies are for example the descriptions about the patients of Freud, who were thoroughly analysed and described.

Bell (1999) states "a case study approach is particularly appropriate for individual researchers because it gives an opportunity for one aspect of a problem to be studied in some depth within a limited time scale".

Ethnographic Study

This type of research is involved with a group, organization, culture, or community. Normally the researcher shares a lot of time with the group.

Grounded Theory Study

Grounded theory research is a systematic research process that works to develop "a process, and action or an interaction about a substantive topic".

References

- Adèr, H. J., Mellenbergh, G. J., & Hand, D. J. (2008). Advising on research methods: a consultant's companion. Huizen: Johannes van Kessel Publishing. ISBN 978-90-79418-01-5
- Muaz, Jalil Mohammad (2013), Practical Guidelines for conducting research. Summarizing good research practice in line with the DCED Standard
- Creswell, J.W. (2012). Educational research: Planning, conducting, and evaluating quantitative and qualitative research. Upper Saddle River, NJ: Prentice Hall.

2

Types of Research Design

Case study is the study that concentrates on a particular person, group or situation. They can be written in the form of journals or presented in professional conferences. Some other types of research design included in this chapter are naturalistic observation, human subject research, causal research, correlation and dependence and scientific method. Research design is best understood in confluence with the major topics listed in the following chapter.

Case Study

A case study is about a person, group, or situation that has been studied over time. If the case study, for instance, is about a group, it describes the behavior of the group as a whole, not the behavior of each individual in the group.

Case studies can be produced by following a formal research method. These case studies are likely to appear in formal research venues, as journals and professional conferences, rather than popular works. The resulting body of 'case study research' has long had a prominent place in many disciplines and professions, ranging from psychology, anthropology, sociology, and political science to education, clinical science, social work, and administrative science.

In doing case study research, the "case" being studied may be an individual, organization, event, or action, existing in a specific time and place. For instance, clinical science has produced both well-known case studies of individuals and also case studies of clinical practices. However, when "case" is used in an abstract sense, as in a claim, a proposition, or an argument, such a case can be the subject of many research methods, not just case study research.

Thomas offers the following definition of case study:

"Case studies are analyses of persons, events, decisions, periods, projects, policies, institutions, or other systems that are studied holistically by one or more method. The case that is the *subject* of the inquiry will be an instance of a class of phenomena that provides an analytical frame — an *object* — within which the study is conducted and which the case illuminates and explicates."

According to J. Creswell, data collection in a case study occurs over a "sustained period of time."

One approach sees the *case study* defined as a *research strategy*, an empirical inquiry that investigates a phenomenon within its real-life context. Case-study research can mean single and multiple case studies, can include quantitative evidence, relies on multiple sources of evidence, and benefits from the prior development of theoretical propositions. As such, case study research should not be confused with qualitative research, as case studies can be based on any mix of quantitative and qualitative data. Similarly, single-subject research might be taken as case studies of a sort, except that the repeated trials in single-subject research permit the use of experimental designs that would not be possible in typical case studies. At the same time, the repeated trials can provide a statistical framework for making inferences from quantitative data.

The case study is sometimes mistaken for the case method used in teaching, but the two are not the same.

Case Selection and Structure

An average, or typical case, is often not the richest in information. In clarifying lines of history and causation it is more useful to select subjects that offer an interesting, unusual or particularly revealing set of circumstances. A case selection that is based on representativeness will seldom be able to produce these kinds of insights. When selecting a subject for a case study, researchers will therefore use information-oriented sampling, as opposed to random sampling. Outlier cases (that is, those which are extreme, deviant or atypical) reveal more information than the potentially representative case. Alternatively, a case may be selected as a key case, chosen because of the inherent interest of the case or the circumstances surrounding it. Alternatively it may be chosen because of a researchers' in-depth local knowledge; where researchers have this local knowledge they are in a position to "soak and poke" as Fenno puts it, and thereby to offer reasoned lines of explanation based on this rich knowledge of setting and circumstances.

Three types of cases may thus be distinguished for selection:

1. Key cases
2. Outlier cases
3. Local knowledge cases

Whatever the frame of reference for the choice of the subject of the case study (key, outlier, local knowledge), there is a distinction to be made between the *subjestorical unity* through which the theoretical focus of the study is being viewed. The object is that theoretical focus – the analytical frame. Thus, for example, if a researcher were interested in US resistance to communist expansion as a theoretical focus, then the Korean War might be taken to be the *subject*, the lens, the case study through which the theoretical focus, the *object*, could be viewed and explicated.

Beyond decisions about case selection and the subject and object of the study, decisions need to be made about purpose, approach and process in the case study. Thomas thus proposes a typology for the case study wherein purposes are first identified (evaluative or exploratory), then approaches are delineated (theory-testing, theory-building or illustrative), then processes are decided upon, with a principal choice being between whether the study is to be single or multiple, and choices also about whether the study is to be retrospective, snapshot or diachronic, and whether it is nested, parallel or sequential. It is thus possible to take many routes through this typology, with, for example, an exploratory, theory-building, multiple, nested study, or an evaluative, theory-testing, single, retrospective study. The typology thus offers many permutations for case-study structure.

A closely related study in medicine is the case report, which identifies a specific case as treated and/or examined by the authors as presented in a novel form. These are, to a differentiable degree, similar to the case study in that many contain reviews of the relevant literature of the topic discussed in the thorough examination of an array of cases published to fit the criterion of the report being presented. These case reports can be thought of as brief case studies with a principal discussion of the new, presented case at hand that presents a novel interest.

Types of Case Studies

Don W. Stacks identifies three types of case study as used in public-relations research:

1. Linear,
2. Process-oriented,
3. Grounded.

Under the more generalized category of case study exist several subdivisions, each of which is custom selected for use depending upon the goals and/or objectives of the investigator. These types of case study include the following:

- Illustrative case studies. These are primarily descriptive studies. They typically utilize one or two instances of an event to show the existing situation. Illustrative case studies serve primarily to make the unfamiliar familiar and to give readers a common language about the topic in question.
- Exploratory (or pilot) case studies. These are condensed case studies performed before implementing a large scale investigation. Their basic function is to help identify questions and select types of measurement prior to the main investigation. The primary pitfall of this type of study is that initial findings may seem convincing enough to be released prematurely as conclusions.
- Cumulative case studies. These serve to aggregate information from several

sites collected at different times. The idea behind these studies is the collection of past studies will allow for greater generalization without additional cost or time being expended on new, possibly repetitive studies.

- Critical instance case studies. These examine one or more sites for either the purpose of examining a situation of unique interest with little to no interest in generalization, or to call into question or challenge a highly generalized or universal assertion. This method is useful for answering cause and effect questions.

Generalizing from Case Studies

A critical case is defined as having strategic importance in relation to the general problem. A critical case allows the following type of generalization: "If it is valid for this case, it is valid for all (or many) cases." In its negative form, the generalization would run: "If it is not valid for this case, then it is not valid for any (or valid for only few) cases."

The case study is effective for generalizing using the type of test that Karl Popper called falsification, which forms part of critical reflexivity. Falsification offers one of the most rigorous tests to which a scientific proposition can be subjected: if just one observation does not fit with the proposition it is considered not valid generally and must therefore be either revised or rejected. Popper himself used the now famous example: "All swans are white", and proposed that just one observation of a single black swan would falsify this proposition and in this way have general significance and stimulate further investigations and theory-building. The case study is well suited for identifying "black swans" because of its in-depth approach: what appears to be "white" often turns out on closer examination to be "black".

Galileo Galilei built his rejection of Aristotle's law of gravity on a case study selected by information-oriented sampling and not by random sampling. The rejection consisted primarily of a conceptual experiment and later on of a practical one. These experiments, with the benefit of hindsight, are self-evident. Nevertheless, Aristotle's incorrect view of gravity had dominated scientific inquiry for nearly two thousand years before it was falsified. In his experimental thinking, Galileo reasoned as follows: if two objects with the same weight are released from the same height at the same time, they will hit the ground simultaneously, having fallen at the same speed. If the two objects are then stuck together into one, this object will have double the weight and will according to the Aristotelian view therefore fall faster than the two individual objects. This conclusion seemed contradictory to Galileo.

History

It is generally believed that the case-study method was first introduced into social science by Frederic Le Play in 1829 as a handmaiden to statistics in his studies of family budgets.

Other roots stem from the early 20th century, when case studies began taking place in the disciplines of sociology, psychology, and anthropology. In all these disciplines, case studies were an occasion for creating new theory, as in the Grounded Theory work of sociologists Barney Glaser and Anselm Strauss.

The popularity of case studies in testing theory or hypotheses has developed only in recent decades. One of the areas in which case studies have been gaining popularity is education and in particular educational evaluation.

Case studies have also been used as a teaching method and as part of professional development, especially in business and legal education. The problem-based learning (PBL) movement is such an example. When used in (non-business) education and professional development, case studies are often referred to as *critical incidents*.

Ethnography is an example of a type of case study, commonly found in communication case studies. Ethnography is the description, interpretation, and analysis of a culture or social group, through field research in the natural environment of the group being studied. The main method of ethnographic research is through observation where the researcher observes the participants over an extended period of time within the participants own environment.

Related Uses

Using case studies in research differs from their use in teaching. At the same time, many people's first exposure to case studies occurred in the classroom, and teaching case studies have been a highly popular pedagogical format in many fields — ranging from business education to science education.

The Harvard Business School has possibly been the most prominent developer and user of teaching case studies. Business school faculty generally develop case studies with particular learning objectives in mind, and the classroom experiences may lead to refinement prior to publication. Additional relevant documentation (such as financial statements, time-lines, and short biographies, often referred to in the case study as "exhibits"), multimedia supplements (such as video-recordings of interviews with the case protagonist), and a carefully crafted teaching note often accompany the case studies. Similarly, teaching case studies have become increasingly popular in science education. The National Center for Case Studies in Teaching Science has made a growing body of case studies available for classroom use, for university as well as secondary school coursework. A new generation of scholars and educators have recently started to call for a more embodied engagement with case studies, using dramaturgy, creative techniques, and emotional involvement, too.

Nevertheless, the principles in doing case study research contrast strongly with those in doing case studies for teaching. The teaching case studies need not adhere strictly to the use of evidence, as they can be manipulated to satisfy pedagogical needs. The generalizations from teaching case studies also may relate to pedagogical issues rather than the substance of the case being studied. Unfortunately, the contrast between the two types of case studies have not always been appreciated. For this reason, many people have had poor impressions of the validity and generalizability of case study research.

Case studies are commonly used in case competitions and interviews for consulting firms such as McKinsey & Company, CEB Inc. and the Boston Consulting Group, in which candidates are asked to develop the best solution for a case in an allotted time frame.

Naturalistic Observation

Naturalistic observation is, in contrast to analog observation, a research tool in which a subject is observed in its natural habitat without any manipulation by the observer. During naturalistic observation, researchers take great care to avoid interfering with the behavior they are observing by using unobtrusive methods. Naturalistic observation involves two main differences that set it apart from other forms of data gathering. In the context of a naturalistic observation the environment is in no way being manipulated by the observer nor was it created by the observer.

Naturalistic observation, as a research tool, comes with both advantages and disadvantages that impact its application. By merely observing at a given instance without any manipulation in its natural context, it makes the behaviors exhibited more credible because they are occurring in a real, typical scenario as opposed to an artificial one generated within a lab. Naturalistic observation also allows for study of events that are deemed unethical to study via experimental models, such as the impact of high school shootings on students attending the high school. Naturalistic observation is used in many techniques, from watching an animals eating patterns in the forest to observing the behavior of students in a school setting.

Human Subject Research

1946 military human subject research on the effects of wind on humans

Human subject research is systematic, scientific investigation that can be either interventional (a "trial") or observational and involves human beings as research subjects. Human subject research can be either medical (clinical) research or non-medical (e.g., social science) research. Systematic investigation incorporates both the collection and analysis of data in order to answer a specific question. Medical human subject research often involves analysis of biological specimens, epidemiological and behavioral studies and medical chart review studies. (A specific, and especially heavily regulated, type of medical human subject research is the "clinical trial", in which drugs, vaccines and medical devices are evaluated.) On the other hand, human subject research in the social sciences often involves surveys which consist of questions to a particular group of people. Survey methodology includes questionnaires, interviews, and focus groups.

Human subject research is used in various fields, including research into basic biology, clinical medicine, nursing, psychology, sociology, political science, and anthropology. As research has become formalized, the academic community has developed formal definitions of "human subject research", largely in response to abuses of human subjects.

Human Subjects

The United States Department of Health and Human Services (HHS) defines a human research subject as a living individual about whom a research investigator (whether a professional or a student) obtains data through 1) intervention or interaction with the individual, or 2) identifiable private information (32 C.F.R. 219.102(f)). (Lim, 1990)

As defined by HHS regulations:

"Intervention"- physical procedures by which data is gathered and the manipulation of the subject and/or their environment for research purposes [45 C.F.R. 46.102(f)]

"Interaction"- communication or interpersonal contact between investigator and subject [45 C.F.R. 46.102(f)])

"Private Information"- information about behavior that occurs in a context in which an individual can reasonably expect that no observation or recording is taking place, and information which has been provided for specific purposes by an individual and which the individual can reasonably expect will not be made public [45 C.F.R. 46.102(f)])]

"Identifiable information"- specific information that can be used to identify an individual

Human Subject Rights

In 2010, the National Institute of Justice in the United States published recommended rights of human subjects:

- Voluntary, informed consent
- Respect for persons: treated as autonomous agents
- The right to end participation in research at any time
- Right to safeguard integrity
- Benefits should outweigh cost
- Protection from physical, mental and emotional harm
- Access to information regarding research
- Protection of privacy and well-being

Ethical Guidelines

Ethical guidelines that govern the use of human subjects in research are a fairly new construct. In 1906 some regulations were put in place in the United States to protect subjects from abuses. After the passage of the Pure Food and Drug Act in 1906, regulatory bodies were gradually institutionalized such as the Food and Drug Administration (FDA) and the Institutional Review Board (IRB). The policies that these institutions implemented served to minimize harm to the participant's mental and/or physical well being.

Nuremberg Code

In 1947, German physicians who conducted deadly or debilitating experiments on concentration camp prisoners were prosecuted as war criminals in the Nuremberg Trials. That same year, the Allies established the Nuremberg Code, the first international document to support the concept that "the voluntary consent of the human subject is absolutely essential". Individual consent was emphasized in the Nuremberg Code in order to prevent prisoners of war, patients, prisoners, and soldiers from being coerced into becoming human subjects. In addition, it was emphasized in order to inform participants of the risk-benefit outcomes of experiments.

Declaration of Helsinki

The Declaration of Helsinki was established in 1964 to regulate international research involving human subjects. Established by the World Medical Association, the declaration recommended guidelines for medical doctors conducting biomedical research that involves human subjects. Some of these guidelines included the principles that "research protocols should be reviewed by an independent committee prior to initiation" and that "research with humans should be based on results from laboratory animals and experimentation".

The Declaration of Helsinki is widely regarded as the cornerstone document on human research ethics.

Clinical Trials

Clinical trials are experiments done in clinical research. Such prospective biomedical or behavioral research studies on human participants are designed to answer specific questions about biomedical or behavioral interventions, including new treatments (such as novel vaccines, drugs, dietary choices, dietary supplements, and medical devices) and known interventions that warrant further study and comparison. Clinical trials generate data on safety and efficacy. They are conducted only after they have received health authority/ethics committee approval in the country where approval of the therapy is sought. These authorities are responsible for vetting the risk/benefit ratio of the trial - their approval does not mean that the therapy is 'safe' or effective, only that the trial may be conducted.

Depending on product type and development stage, investigators initially enroll volunteers and/or patients into small pilot studies, and subsequently conduct progressively larger scale comparative studies. Clinical trials can vary in size and cost, and they can involve a single research center or multiple centers, in one country or in multiple countries. Clinical study design aims to ensure the scientific validity and reproducibility of the results.

Trials can be quite costly, depending on a number of factors. The sponsor may be a governmental organization or a pharmaceutical, biotechnology or medical device company. Certain functions necessary to the trial, such as monitoring and lab work, may be managed by an outsourced partner, such as a contract research organization or a central laboratory.

Human Subjects in Psychology and Sociology

Stanford Prison Experiment

A study conducted by Philip Zimbardo in 1971 examined the effect of social roles on college students at Stanford University. Twenty-four male students were assigned to a random role of a prisoner or guard to simulate a mock prison in one of Stanford's basements. After only six days, the abusive behavior of the guards and the psychological suffering of prisoners proved significant enough to halt the two-week-long experiment.

Milgram Experiment

In 1961, Yale University psychologist Stanley Milgram led a series of experiments to determine to what extent an individual would obey instructions given by an experimenter. Placed in a room with the experimenter, subjects played the role of a "teacher" to a "learner" situated in a separate room. The subjects were instructed to administer an electric shock to the learner when the learner answered incorrectly to a set of questions. The intensity of this electric shock was to be increased for every incorrect answer. The learner was a confederate (i.e. actor), and the shocks were faked, but the subjects were led to believe otherwise. Both prerecorded sounds of electric shocks and the confederate's pleas for the punishment to stop were audible to the "teacher" throughout the experiment. When the subject raised questions or paused, the experimenter insisted that the experiment should continue. Despite widespread speculation that most participants would not continue to "shock" the learner, 65 percent of participants in Milgram's initial trial complied until the end of the experiment, continuing to administer shocks to the confederate with purported intensities of up to "450 volts". Although many participants questioned the experimenter and displayed various signs of discomfort, when the experiment was repeated, 65 percent of subjects were willing to obey instructions to administer the shocks through the final one.

Asch Conformity Experiments

Psychologist Solomon Asch's classic conformity experiment in 1951 involved one subject participant and multiple confederates; they were asked to provide answers to a variety of different low-difficulty questions. In every scenario, the multiple confederates gave their answers in turn, and the subject participant subject was allowed to answer last. In a control group of participants, the percentage of error was less than one percent. However, when the confederates unanimously chose an incorrect answer, 75 percent of the subject participants agreed with the majority at least once. The study has been regarded as significant evidence for the power of social influence and conformity.

Robber's Cave Study

A classic advocate of Realistic conflict theory, Muzafer Sherif's Robber's Cave experiment shed light on how group competition can foster hostility and prejudice. In the 1961 study, two groups of ten boys each who were not "naturally" hostile were grouped together without knowledge of one another in Robber's Cave State Park, Oklahoma. The twelve-year-old boys bonded with their own groups for a week before the groups were set in competition with each other in games such as tug-of-war and football. In light of this competition, the groups resorted to name-calling and other displays of resentment, such as burning the other group's team flag. The hostility continued and worsened until the end of the three-week study, when the groups were forced to work together to solve problems.

Bystander Effect

The bystander effect is demonstrated in a series of famous experiments by Bibb Latane and John Darley In each of these experiments, participants were confronted with a type of emergency, such as the witnessing of a seizure or smoke entering through air vents. A common phenomenon was observed that as the number of witnesses or "bystanders" increases, so does the time it takes for individuals to respond to the emergency. This effect has been shown to promote the diffusion of responsibility by concluding that, when surrounded by others, the individual expects someone else to take action.

Cognitive Dissonance

Human subjects have been commonly used in experiments testing the theory of cognitive dissonance after the landmark study by Leon Festinger and Merrill Carlsmith. In 1959, Festinger and Carlsmith devised a situation in which participants would undergo excessively tedious and monotonous tasks. After the completion of these tasks, the subjects were instructed to help the experiment continue in exchange for a variable amount of money. All the subjects had to do was simply inform the next "student" waiting outside the testing area (who was secretly a confederate) that the tasks involved in the experiment were interesting and enjoyable. It was expected that the participants wouldn't fully agree with the information they were imparting to the student, and after complying, half of the participants were awarded $1, and the others were awarded $20. A subsequent survey showed that, by a large margin, those who received less money for essentially "lying" to the student came to believe that the tasks were far more enjoyable than their highly paid counterparts.

Unethical Human Experimentation

Unethical human experimentation violates the principles of medical ethics. It has been performed by countries including Nazi Germany, Imperial Japan, North Korea, the United States, and the Soviet Union. Examples include Project MKUltra, Unit 731, Totskoye nuclear exercise, the experiments of Josef Mengele, and the human experimentation conducted by Chester M. Southam.

Nazi Germany performed human experimentation on large numbers of prisoners (including children), largely Jews from across Europe, but also Romani, Sinti, ethnic Poles, Soviet POWs and disabled Germans, by Nazi Germany in its concentration camps mainly in the early 1940s, during World War II and the Holocaust. Prisoners were forced into participating; they did not willingly volunteer and no consent was given for the procedures. Typically, the experiments resulted in death, trauma, disfigurement or permanent disability, and as such are considered as examples of medical torture. After the war, these crimes were tried at what became known as the Doctors' Trial, and the abuses perpetrated led to the development of the Nuremberg Code. During the Nuremberg Trials, 23 Nazi doctors and scientists were prosecuted for the unethical treatment of concentration camp inmates, who were often used as research subjects with fatal consequences. Of those 23, 15 were convicted, 7 were condemned to death, 9 received prison sentences from 10 years to life, and 7 were acquitted.

Unit 731, a department of the Imperial Japanese Army located near Harbin (then in the puppet state of Manchukuo, in northeast China), experimented on prisoners by conducting vivisections, dismemberments, and bacterial inoculations. It induced epidemics on a very large scale from 1932 onward through the Second Sino-Japanese war. It also conducted biological and chemical weapons tests on prisoners and captured POWs. With the expansion of the empire during World War II, similar units were set up in conquered cities such as Nanking (Unit 1644), Beijing (Unit 1855), Guangzhou (Unit 8604) and Singapore (Unit 9420). After the war, Supreme Commander of the Occupation Douglas MacArthur gave immunity in the name of the United States to Shiro Ishii and all members of the units in exchange for all of the results of their experiments.

During World War II, Fort Detrick in Maryland was the headquarters of US biological warfare experiments. Operation Whitecoat involved the injection of infectious agents into military forces to observe their effects in human subjects. Subsequent human experiments in the United States have also been characterized as unethical. They were often performed illegally, without the knowledge, consent, or informed consent of the test subjects. Public outcry over the discovery of government experiments on human subjects led to numerous congressional investigations and hearings, including the Church Committee, Rockefeller Commission, and Advisory Committee on Human Radiation Experiments, amongst others. The Tuskegee syphilis experiment, widely regarded as the "most infamous biomedical research study in U.S. history," was performed from 1932 to 1972 by the Tuskegee Institute contracted by the United States Public Health Service. The study followed more than 600 African-American men who were not told they had syphilis and were denied access to the known treatment of penicillin. This led to the 1974 National Research Act, to provide for protection of human subjects in experiments. The National Commission for the Protection of Human Subjects of Biomedical and Behavioral Research was established and was tasked with establishing the boundary between research and routine practice, the role of risk-benefit analysis, guidelines for participation, and the definition of informed consent. Its *Belmont Re-*

port established three tenets of ethical research: respect for persons, beneficence, and justice.

From the 1950's-60's, Chester M. Southam, an important virologist and cancer researcher, injected HeLa cells into cancer patients, healthy individuals, and prison inmates from the Ohio Penitentiary. He wanted to observe if cancer could be transmitted as well as if people could become immune to cancer by developing an acquired immune response. Many believe that this experiment violated the bioethical principles of informed consent, non-maleficence, and beneficence .

Survey (Human Research)

In research of human subjects, a survey is a list of questions aimed at extracting specific data from a particular group of people. Surveys may be conducted by phone, mail, via the internet, and sometimes face-to-face on busy street corners or in malls. Surveys are used to increase knowledge in fields such as social research and demography.

Survey research is often used to assess thoughts, opinions, and feelings. Surveys can be specific and limited, or they can have more global, widespread goals. Psychologists and sociologists often use surveys to analyze behavior, while it is also used to meet the more pragmatic needs of the media, such as, in evaluating political candidates, public health officials, professional organizations, and advertising and marketing directors. A survey consists of a predetermined set of questions that is given to a sample. With a representative sample, that is, one that is representative of the larger population of interest, one can describe the attitudes of the population from which the sample was drawn. Further, one can compare the attitudes of different populations as well as look for changes in attitudes over time. A good sample selection is key as it allows one to generalize the findings from the sample to the population, which is the whole purpose of survey research.

Types

Census

A *census* is the procedure of systematically acquiring and recording information about the members of a given population. It is a regularly occurring and official count of a particular population. The term is used mostly in connection with national population and housing censuses; other common censuses include agriculture, business, and traffic censuses. The United Nations defines the essential features of population and housing censuses as "individual enumeration, universality within a defined territory, simultaneity and defined periodicity", and recommends that population censuses be taken at least every 10 years.

Other Household Surveys

Other surveys than the census may explore characteristics in households, such as fertility, family structure, and demographics.

Household surveys with at least 10,000 participants include:

- General Household Survey, conducted in private households in Great Britain. It is a repeated cross-sectional study, conducted annually, which uses a sample of 9,731 households in the 2006 survey.
- Generations and Gender Survey, conducted in several countries in Europe as well as Australia and Japan. The programme has collected least one wave of surveys in 19 countries, with an average of 9,000 respondents per country.
- Household, Income and Labour Dynamics in Australia Survey, where the wave 1 panel consisted of 7,682 households and 19,914 individuals
- Integrated Household Survey, a survey made up of multiple other surveys in the UK. It includes about 340,000 respondents, making it the largest collection of social data in the UK after the census.
- National Survey of Family Growth, conducted in the United States by the National Center for Health Statistics division of the Centers for Disease Control and Prevention to understand trends related to fertility, family structure, and demographics in the United States. The 2006-2010 NSFG surveyed 22,682 interviews.
- Panel Study of Income Dynamics in the United States, wherein data have been collected from the same families and their descendants since 1968. The study started with over 18,000 nationally representative individuals. It involved more than 9,000 individuals as of 2009.
- Socio-Economic Panel, a longitudinal panel dataset of the population in Germany. It is a household based study which started in 1984 and which reinterviews adult household members annually. In 2007, the study involved about 12,000 households, with more than 20,000 adult persons sampled.
- UK households: a longitudinal study, now known as *Understanding Society*. Its sample size is 40,000 households from the United Kingdom or approx. 100,000 individuals.

Opinion Poll

An *opinion poll* is a survey of public opinion from a particular sample. Opinion polls are usually designed to represent the opinions of a population by conducting a series of questions and then extrapolating generalities in ratio or within confidence intervals.

November 3, 1948: President Harry S. Truman, shortly after being elected as President, smiles as he holds up a copy of the *Chicago Tribune* issue predicting his electoral defeat. This image has become iconic of the consequences of bad polling data.

Methodology

A single survey is made of at least a sample (or full population in the case of a census), a method of data collection (e.g., a questionnaire) and individual questions or items that become data that can be analyzed statistically. A single survey may focus on different types of topics such as preferences (e.g., for a presidential candidate), opinions (e.g., should abortion be legal?), behavior (smoking and alcohol use), or factual information (e.g., income), depending on its purpose. Since survey research is almost always based on a sample of the population, the success of the research is dependent on the representativeness of the sample with respect to a target population of interest to the researcher. That target population can range from the general population of a given country to specific groups of people within that country, to a membership list of a professional organization, or list of students enrolled in a school system.

Interpretation

Correlation and Causality

When two variables are related, or correlated, one can make predictions for these two variables. However, it is important to note that this does not mean causality. At this point, it is not possible to determine a causal relationship between the two variables; correlation does not imply causality. However, correlation evidence is significant because it can help identify potential causes of behavior. Path analysis is a statistical technique that can be used with correlational data. This involves the identification of mediator and moderator variables. A mediator variable is used to explain the correlation between two variables. A moderator variable affects the direction or strength of the correlation between two variables. A spurious relationship is a relationship in which the relation between two variables can be explained by a third variable.

Reported Behavior Versus Actual Behavior

The value of collected data completely depends upon how truthful respondents are in their answers on questionnaires. In general, survey researchers accept respondents' answers as true. Survey researchers avoid reactive measurement by examining the accuracy of verbal reports, and directly observing respondents' behavior in comparison with their verbal reports to determine what behaviors they really engage in or what attitudes they really uphold. Studies examining the association between self reports (attitudes, intentions) and actual behavior show that the link between them—though positive—is not always strong—thus caution is needed when extrapolating self-reports to actual behaviors,

History

The most famous public survey in America is the national census. Held every ten years, the census attempts to count all persons, and also obtain demographic data about factors such as household income, ethnicity, and religion. The most recent survey was conducted in 2010.

Nielsen ratings are another example of public surveys. Nielsen ratings track media-viewing habits (radio, television, internet, print) the results of which are used to make decisions by and about the mass media. Some Nielsen ratings localize the data points to give marketing firms more specific information with which to target customers. Demographic data is also used to understand what influences work best to market consumer products, political campaigns, etc.

Quasi-Experiment

A quasi-experiment is an empirical study used to estimate the causal impact of an intervention on its target population without random assignment. Quasi-experimental research shares similarities with the traditional experimental design or randomized controlled trial, but they specifically lack the element of random assignment to treatment or control. Instead, quasi-experimental designs typically allow the researcher to control the assignment to the treatment condition, but using some criterion other than random assignment (e.g., an eligibility cutoff mark). In some cases, the researcher may have control over assignment to treatment. Quasi-experiments are subject to concerns regarding internal validity, because the treatment and control groups may not be comparable at baseline. With random assignment, study participants have the same chance of being assigned to the intervention group or the comparison group. As a result, differences between groups on both observed and unobserved characteristics would be due to chance, rather than to a systematic factor related to treatment (e.g., illness severity).

Randomization itself does not guarantee that groups will be equivalent at baseline. Any change in characteristics post-intervention is likely attributable to the intervention. With quasi-experimental studies, it may not be possible to convincingly demonstrate a causal link between the treatment condition and observed outcomes. This is particularly true if there are confounding variables that cannot be controlled or accounted for.

Design

The first part of creating a quasi-experimental design is to identify the variables. The quasi-independent variable will be the x-variable, the variable that is manipulated in order to affect a dependent variable. "X" is generally a grouping variable with different levels. Grouping means two or more groups, such as two groups receiving alternative treatments, or a treatment group and a no-treatment group (which may be given a placebo - placebos are more frequently used in medical or physiological experiments). The predicted outcome is the dependent variable, which is the y-variable. In a time series analysis, the dependent variable is observed over time for any changes that may take place. Once the variables have been identified and defined, a procedure should then be implemented and group differences should be examined.

In an experiment with random assignment, study units have the same chance of being assigned to a given treatment condition. As such, random assignment ensures that both the experimental and control groups are equivalent. In a quasi-experimental design, assignment to a given treatment condition is based on something other than random assignment. Depending on the type of quasi-experimental design, the researcher might have control over assignment to the treatment condition but use some criteria other than random assignment (e.g., a cutoff score) to determine which participants receive the treatment, or the researcher may have no control over the treatment condition assignment and the criteria used for assignment may be unknown. Factors such as cost, feasibility, political concerns, or convenience may influence how or if participants are assigned to a given treatment conditions, and as such, quasi-experiments are subject to concerns regarding internal validity (i.e., can the results of the experiment be used to make a causal inference?).

Quasi-experiments are also effective because they use the "pre-post testing". This means that there are tests done before any data is collected to see if there are any person confounds or if any participants have certain tendencies. Then the actual experiment is done with post test results recorded. This data can be compared as part of the study or the pre-test data can be included in an explanation for the actual experimental data. Quasi experiments have independent variables that already exist such as age, gender, eye color. These variables can either be continuous (age) or they can be categorical (gender). In short, naturally occurring variables are measured within quasi experiments.

There are several types of quasi-experimental designs, each with different strengths, weaknesses and applications. These designs include (but are not limited to):

- Difference in differences (pre-post with-without comparison)
- Nonequivalent control groups design
    - no-treatment control group designs
    - nonequivalent dependent variables designs
    - removed treatment group designs
    - repeated treatment designs
    - reversed treatment nonequivalent control groups designs
    - cohort designs
    - post-test only designs
    - regression continuity designs
- Regression discontinuity design
- case-control design
    - time-series designs
    - multiple time series design
    - interrupted time series design
    - Propensity score matching
    - Instrumental variables
- Panel analysis

Of all of these designs, the regression discontinuity design comes the closest to the experimental design, as the experimenter maintains control of the treatment assignment and it is known to "yield an unbiased estimate of the treatment effects". It does, however, require large numbers of study participants and precise modeling of the functional form between the assignment and the outcome variable, in order to yield the same power as a traditional experimental design.

Though quasi-experiments are sometimes shunned by those who consider themselves to be experimental purists (leading Donald T. Campbell to coin the term "queasy experiments" for them), they are exceptionally useful in areas where it is not feasible or desirable to conduct an experiment or randomized control trial. Such instances include evaluating the impact of public policy changes, educational interventions or large scale health interventions. The primary drawback of quasi-experimental designs is that they cannot eliminate the possibility of confounding bias, which can hinder one's ability to draw causal inferences. This drawback is often used to discount quasi-experimental results. However, such bias can be controlled for using various statistical techniques such as multiple regression, if one can identify and measure the confounding variable(s). Such techniques can be used to model and partial out the effects of confounding variables techniques, thereby improving the accuracy of the results obtained from quasi-experiments. Moreover, the developing use of propensity score matching to match participants on variables important to the treatment selection process can also improve the accuracy of quasi-experimental results. In sum, quasi-experiments are a valuable tool, especially for the applied researcher. On their own, quasi-experimental designs do not allow one to make definitive causal inferences; however, they provide necessary and valuable information that cannot be obtained by experimental methods alone. Researchers, especially those interested in investigating applied research questions, should move beyond the traditional experimental design and avail themselves of the possibilities inherent in quasi-experimental designs.

Ethics

A true experiment would randomly assign children to a scholarship, in order to control for all other variables. Quasi-experiments are commonly used in social sciences, public health, education, and policy analysis, especially when it is not practical or reasonable to randomize study participants to the treatment condition.

As an example, suppose we divide households into two categories: Households in which the parents spank their children, and households in which the parents do not spank their children. We can run a linear regression to determine if there is a positive correlation between parents' spanking and their children's aggressive behavior. However, to simply randomize parents to spank or to not spank their children may not be practical or ethical, because some parents may believe it is morally wrong to spank their children and refuse to participate.

Some authors distinguish between a natural experiment and a "quasi-experiment". The difference is that in a quasi-experiment the criterion for assignment is selected by the researcher, while in a natural experiment the assignment occurs 'naturally,' without the researcher's intervention.

Quasi experiments have outcome measures, treatments, and experimental units, but do not use random assignment. Quasi-experiments are often the design that most people choose over true experiments. The main reason is that they can usually be conducted while true experiments can not always be. Quasi-experiments are interesting because they bring in features from both experimental and non experimental designs. Measured variables can be brought in, as well as manipulated variables. Usually Quasi-experiments are chosen by experimenters because they maximize internal and external validity.

Advantages

Since quasi-experimental designs are used when randomization is impractical and/ or unethical, they are typically easier to set up than true experimental designs, which require random assignment of subjects. Additionally, utilizing quasi-experimental designs minimizes threats to ecological validity as natural environments do not suffer the same problems of artificiality as compared to a well-controlled laboratory setting. Since quasi-experiments are natural experiments, findings in one may be applied to other subjects and settings, allowing for some generalizations to be made about population. Also, this experimentation method is efficient in longitudinal research that involves longer time periods which can be followed up in different environments.

Other advantages of quasi experiments include the idea of having any manipulations the experimenter so chooses. In natural experiments, the researchers have to let manipulations occur on their own and have no control over them whatsoever. Also, using self selected groups in quasi experiments also takes away to chance of ethical, conditional, etc. concerns while conducting the study.

Disadvantages

Quasi-experimental estimates of impact are subject to contamination by confounding variables. In the example above, a variation in the children's response to spanking is plausibly influenced by factors that cannot be easily measured and controlled, for example the child's intrinsic wildness or the parent's irritability. The lack of random assignment in the quasi-experimental design method may allow studies to be more feasible, but this also poses many challenges for the investigator in terms of internal validity. This deficiency in randomization makes it harder to rule out confounding variables and introduces new threats to internal validity. Because randomization is absent, some knowledge about the data can be approximated, but conclusions of causal relationships are difficult to determine due to a variety of extraneous and confounding variables that exist in a social environment. Moreover, even if these threats to internal validity are assessed, causation still cannot be fully established because the experimenter does not have total control over extraneous variables.

Disadvantages also include the study groups may provide weaker evidence because of the lack of randomness. Randomness brings a lot of useful information to a study because it broadens results and therefore gives a better representation of the population as a whole. Using unequal groups can also be a threat to internal validity. If groups are not equal, which is not always the case in quasi experiments, then the experimenter might not be positive what the causes are for the results.

Internal Validity

Internal validity is the approximate truth about inferences regarding cause-effect or causal relationships. This is why validity is important for quasi experiments because they are all about causal relationships. It occurs when the experimenter tries to control all variables that could affect the results of the experiment. Statistical regression, history and the participants are all possible threats to internal validity. The question you would want to ask while trying to keep internal validity high is "Are there any other possible reasons for the outcome besides the reason I want it to be?" If so, then internal validity might not be as strong.

External Validity

External validity is the extent to which results obtained from a study sample can be generalized to the population of interest. When External Validity is high, the generalization is accurate and can represent the outside world from the experiment. External Validity is very important when it comes to statistical research because you want to make sure that you have a correct depiction of the population. When external validity is low, the credibility of your research comes into doubt. Reducing threats to external validity can be done by making sure there is a random sampling of participants and random assignment as well.

Design Types

"Person-by-treatment" designs are the most common type of quasi experiment design. In this design, the experimenter measures at least one independent variable. Along with measuring one variable, the experimenter will also manipulate a different independent variable. Because there is manipulating and measuring of different independent variables, the research is mostly done in laboratories. An important factor in dealing with person-by-treatment designs are that random assignment will need to be used in order to make sure that the experimenter has complete control over the manipulations that are being done to the study.

An example of this type of design was performed at the University of Notre Dame. The study was conducted to see if being mentored for your job led to increased job satisfaction. The results showed that many people who did have a mentor showed very high job satisfaction. However, the study also showed that those who did not receive the mentor also had a high number of satisfied employees. Seibert concluded that although the workers who had mentors were happy, he could not assume that the reason for it was the mentors themselves because of the numbers of the high number of non-mentored employees that said they were satisfied. This is why prescreening is very important so that you can minimize any flaws in the study before they are seen.

"Natural Experiments" are a different type of quasi experiment design used by researchers. It differs from person-by-treatment in a way that there is not a variable that is being manipulated by the experimenter. Instead of controlling at least one variable like the person-by-treatment design, experimenters do not use random assignment and leave the experimental control up to chance. This is where the name "Natural" Experiment comes from. The manipulations occur naturally, and although this may seem like an inaccurate technique, it has actually proven to be useful in many cases. These are the studies done to people who had something sudden happen to them. This could mean good or bad, traumatic or euphoric. An example of this could be studies done on those who have been in a car accident and those who have not. Car accidents occur naturally, so it would not be ethical to stage experiments to traumatize subjects in the study. These naturally occurring events have proven to be useful for studying post traumatic stress disorder cases.

Causal Research

Causal research, also called explanatory research. is the investigation of (research into) cause-and-effect relationships. To determine causality, it is important to observe variation in the variable assumed to cause the change in the other variable(s), and then measure the changes in the other variable(s). Other confounding influences must be controlled for so they don't distort the results, either by holding them constant in the experimental creation of data, or by using statistical methods. This type of research is very complex and the researcher can never be completely certain that there are no other factors influencing the causal relationship, especially when dealing with people's attitudes and motivations. There are often much deeper psychological considerations that even the respondent may not be aware of.

There are two research methods for exploring the cause-and-effect relationship between variables:

1. Experimentation (e.g., in a laboratory), and
2. Statistical research.

Experimentation

Experiments are typically conducted in laboratories where many or all aspects of the experiment can be tightly controlled to avoid spurious results due to factors other than the hypothesized causative factor(s). Many studies in physics, for example, use this approach. Alternatively, field experiments can be performed, as with medical studies in which subjects may have a great many attributes that cannot be controlled for but in which at least the key hypothesized causative variables can be varied and some of the extraneous attributes can at least be measured. Field experiments also are sometimes used in economics, such as when two different groups of welfare recipients are given two alternative sets of incentives or opportunities to earn income and the resulting effect on their labor supply is investigated.

Statistical Research

In areas such as economics, most empirical research is done on pre-existing data, often collected on a regular basis by a government. Multiple regression is a group of related statistical techniques that control for (attempt to avoid spurious influence from) various causative influences other than the ones being studied. If the data show sufficient variation in the hypothesized explanatory variable of interest, its effect if any upon the potentially influenced variable can be measured.

Correlation and Dependence

In statistics, dependence or association is any statistical relationship, whether causal or not, between two random variables or two sets of data. Correlation is any of a broad class of statistical relationships involving dependence, though in common usage it most often refers to the extent to which two variables have a linear relationship with each other. Familiar examples of dependent phenomena include the correlation between the physical statures of parents and their offspring, and the correlation between the demand for a product and its price.

Correlations are useful because they can indicate a predictive relationship that can be exploited in practice. For example, an electrical utility may produce less power on a mild day based on the correlation between electricity demand and weather. In this example there is a causal relationship, because extreme weather causes people to use more electricity for heating or cooling; however, correlation is not sufficient to demonstrate the presence of such a causal relationship (i.e., correlation does not imply causation).

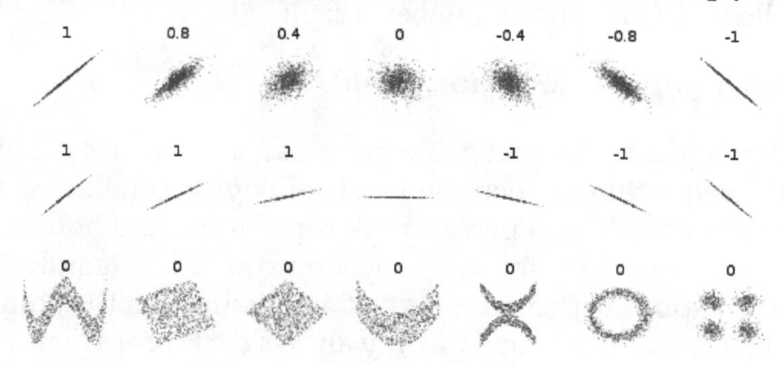

Several sets of (x, y) points, with the Pearson correlation coefficient of x and y for each set. Note that the correlation reflects the noisiness and direction of a linear relationship (top row), but not the slope of that relationship (middle), nor many aspects of nonlinear relationships (bottom). N.B.: the figure in the center has a slope of 0 but in that case the correlation coefficient is undefined because the variance of Y is zero.

Formally, *dependence* refers to any situation in which random variables do not satisfy a mathematical condition of probabilistic independence. In loose usage, *correlation* can refer to any departure of two or more random variables from independence, but technically it refers to any of several more specialized types of relationship between mean values. There are several correlation coefficients, often denoted ρ or r, measuring the degree of correlation. The most common of these is the Pearson correlation coefficient, which is sensitive only to a linear relationship between two variables (which may exist even if one is a nonlinear function of the other). Other correlation coefficients have been developed to be more robust than the Pearson correlation – that is, more sensitive to nonlinear relationships. Mutual information can also be applied to measure dependence between two variables.

Rank Correlation Coefficients

Rank correlation coefficients, such as Spearman's rank correlation coefficient and Kendall's rank correlation coefficient (τ) measure the extent to which, as one variable increases, the other variable tends to increase, without requiring that increase to be represented by a linear relationship. If, as the one variable increases, the other *decreases*, the rank correlation coefficients will be negative. It is common to regard these rank correlation coefficients as alternatives to Pearson's coefficient, used either to reduce the amount of calculation or to make the coefficient less sensitive to non-normality in distributions. However, this view has little mathematical basis, as rank correlation coefficients measure a different type of relationship than the Pearson product-moment correlation coefficient, and are best seen as measures of a different type of association, rather than as alternative measure of the population correlation coefficient.

To illustrate the nature of rank correlation, and its difference from linear correlation, consider the following four pairs of numbers (x, y):

$$(0, 1), (10, 100), (101, 500), (102, 2000).$$

As we go from each pair to the next pair x increases, and so does y. This relationship is perfect, in the sense that an increase in x is *always* accompanied by an increase in y. This means that we have a perfect rank correlation, and both Spearman's and Kendall's correlation coefficients are 1, whereas in this example Pearson product-moment correlation coefficient is 0.7544, indicating that the points are far from lying on a straight line. In the same way if y always *decreases* when x *increases*, the rank correlation coefficients will be −1, while the Pearson product-moment correlation coefficient may or may not be close to −1, depending on how close the points are to a straight line. Although in the extreme cases of perfect rank correlation the two coefficients are both equal (being both +1 or both −1), this is not generally the case, and so values of the two coefficients cannot meaningfully be compared. For example, for the three pairs (1, 1) (2, 3) (3, 2) Spearman's coefficient is 1/2, while Kendall's coefficient is 1/3.

Other Measures of Dependence among Random Variables

The information given by a correlation coefficient is not enough to define the dependence structure between random variables. The correlation coefficient completely defines the dependence structure only in very particular cases, for example when the distribution is a multivariate normal distribution. In the case of elliptical distributions it characterizes the (hyper-)ellipses of equal density, however, it does not completely characterize the dependence structure (for example, a multivariate t-distribution's degrees of freedom determine the level of tail dependence).

Distance correlation was introduced to address the deficiency of Pearson's correlation that it can be zero for dependent random variables; zero distance correlation implies independence.

The Randomized Dependence Coefficient is a computationally efficient, copula-based measure of dependence between multivariate random variables. RDC is invariant with respect to non-linear scalings of random variables, is capable of discovering a wide range of functional association patterns and takes value zero at independence.

The correlation ratio is able to detect almost any functional dependency, and the entropy-based mutual information, total correlation and dual total correlation are capable of detecting even more general dependencies. These are sometimes referred to as multi-moment correlation measures, in comparison to those that consider only second moment (pairwise or quadratic) dependence.

The polychoric correlation is another correlation applied to ordinal data that aims to estimate the correlation between theorised latent variables.

One way to capture a more complete view of dependence structure is to consider a copula between them.

The coefficient of determination generalizes the correlation coefficient for relationships beyond simple linear regression.

Sensitivity to the Data Distribution

The degree of dependence between variables X and Y does not depend on the scale on which the variables are expressed. That is, if we are analyzing the relationship between X and Y, most correlation measures are unaffected by transforming X to $a + bX$ and Y to $c + dY$, where a, b, c, and d are constants (b and d being positive). This is true of some correlation statistics as well as their population analogues. Some correlation statistics, such as the rank correlation coefficient, are also invariant to monotone transformations of the marginal distributions of X and/or Y.

Correlation and Dependence

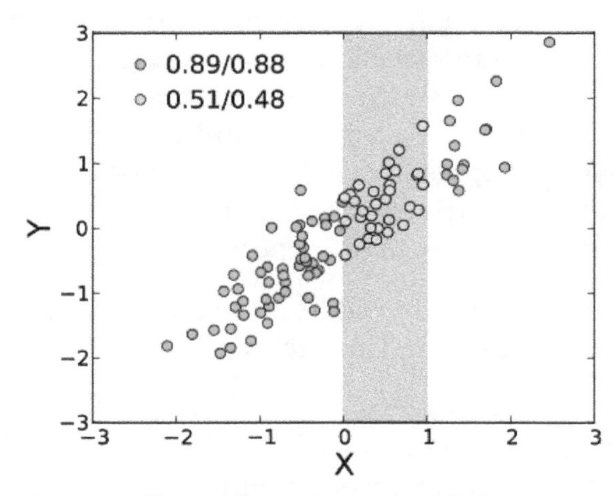

Pearson/Spearman correlation coefficients between X and Y are shown when the two variables' ranges are unrestricted, and when the range of X is restricted to the interval (0,1).

Most correlation measures are sensitive to the manner in which X and Y are sampled. Dependencies tend to be stronger if viewed over a wider range of values. Thus, if we consider the correlation coefficient between the heights of fathers and their sons over all adult males, and compare it to the same correlation coefficient calculated when the fathers are selected to be between 165 cm and 170 cm in height, the correlation will be weaker in the latter case. Several techniques have been developed that attempt to correct for range restriction in one or both variables, and are commonly used in meta-analysis; the most common are Thorndike's case II and case III equations.

Various correlation measures in use may be undefined for certain joint distributions of X and Y. For example, the Pearson correlation coefficient is defined in terms of moments, and hence will be undefined if the moments are undefined. Measures of dependence based on quantiles are always defined. Sample-based statistics intended to estimate population measures of dependence may or may not have desirable statistical properties such as being unbiased, or asymptotically consistent, based on the spatial structure of the population from which the data were sampled.

Sensitivity to the data distribution can be used to an advantage. For example, scaled correlation is designed to use the sensitivity to the range in order to pick out correlations between fast components of time series. By reducing the range of values in a controlled manner, the correlations on long time scale are filtered out and only the correlations on short time scales are revealed.

Correlation Matrices

The correlation matrix of n random variables $X_1, ..., X_n$ is the $n \times n$ matrix whose i,j

entry is corr(X_i, X_j). If the measures of correlation used are product-moment coefficients, the correlation matrix is the same as the covariance matrix of the standardized random variables $X_i / \sigma(X_i)$ for $i = 1, ..., n$. This applies to both the matrix of population correlations (in which case "σ" is the population standard deviation), and to the matrix of sample correlations (in which case "σ" denotes the sample standard deviation). Consequently, each is necessarily a positive-semidefinite matrix.

The correlation matrix is symmetric because the correlation between X_i and X_j is the same as the correlation between X_j and X_i.

Common Misconceptions

Correlation and Causality

The conventional dictum that "correlation does not imply causation" means that correlation cannot be used to infer a causal relationship between the variables. This dictum should not be taken to mean that correlations cannot indicate the potential existence of causal relations. However, the causes underlying the correlation, if any, may be indirect and unknown, and high correlations also overlap with identity relations (tautologies), where no causal process exists. Consequently, establishing a correlation between two variables is not a sufficient condition to establish a causal relationship (in either direction).

A correlation between age and height in children is fairly causally transparent, but a correlation between mood and health in people is less so. Does improved mood lead to improved health, or does good health lead to good mood, or both? Or does some other factor underlie both? In other words, a correlation can be taken as evidence for a possible causal relationship, but cannot indicate what the causal relationship, if any, might be.

Correlation and Linearity

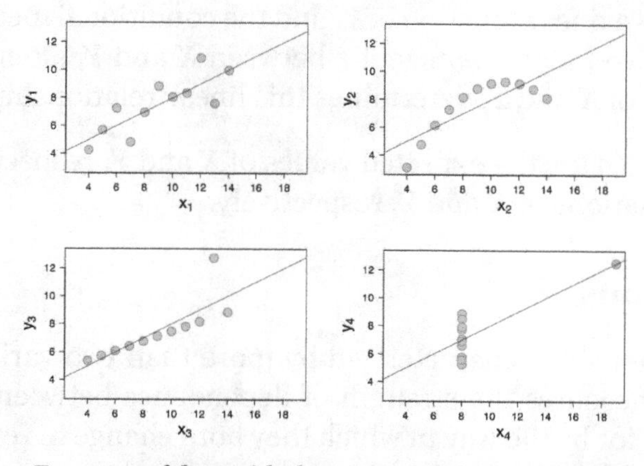

Four sets of data with the same correlation of 0.816

The Pearson correlation coefficient indicates the strength of a *linear* relationship between two variables, but its value generally does not completely characterize their relationship. In particular, if the conditional mean of Y given X, denoted $E(Y \mid X)$, is not linear in X, the correlation coefficient will not fully determine the form of $E(Y \mid X)$.

The image on the right shows scatter plots of Anscombe's quartet, a set of four different pairs of variables created by Francis Anscombe. The four y variables have the same mean (7.5), variance (4.12), correlation (0.816) and regression line ($y = 3 + 0.5x$). However, as can be seen on the plots, the distribution of the variables is very different. The first one (top left) seems to be distributed normally, and corresponds to what one would expect when considering two variables correlated and following the assumption of normality. The second one (top right) is not distributed normally; while an obvious relationship between the two variables can be observed, it is not linear. In this case the Pearson correlation coefficient does not indicate that there is an exact functional relationship: only the extent to which that relationship can be approximated by a linear relationship. In the third case (bottom left), the linear relationship is perfect, except for one outlier which exerts enough influence to lower the correlation coefficient from 1 to 0.816. Finally, the fourth example (bottom right) shows another example when one outlier is enough to produce a high correlation coefficient, even though the relationship between the two variables is not linear.

These examples indicate that the correlation coefficient, as a summary statistic, cannot replace visual examination of the data. Note that the examples are sometimes said to demonstrate that the Pearson correlation assumes that the data follow a normal distribution, but this is not correct.

Bivariate Normal Distribution

If a pair (X, Y) of random variables follows a bivariate normal distribution, the conditional mean $E(X|Y)$ is a linear function of Y, and the conditional mean $E(Y|X)$ is a linear function of X. The correlation coefficient r between X and Y, along with the marginal means and variances of X and Y, determines this linear relationship:

where $E(X)$ and $E(Y)$ are the expected values of X and Y, respectively, and σ_x and σ_y are the standard deviations of X and Y, respectively.

Partial Correlation

If a population or data-set is characterized by more than two variables, a partial correlation coefficient measures the strength of dependence between a pair of variables that is not accounted for by the way in which they both change in response to variations in a selected subset of the other variables.

Scientific Method

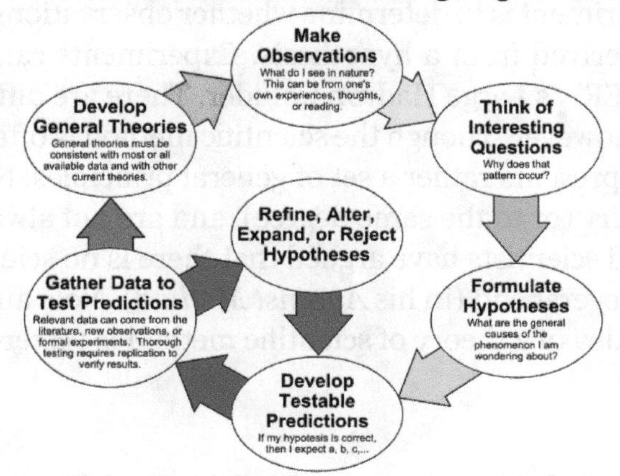

The scientific method as a cyclic or iterative process

The scientific method is a body of techniques for investigating phenomena, acquiring new knowledge, or correcting and integrating previous knowledge. To be termed scientific, a method of inquiry is commonly based on empirical or measurable evidence subject to specific principles of reasoning. The Oxford Dictionaries Online define the scientific method as "a method or procedure that has characterized natural science since the 17th century, consisting in systematic observation, measurement, and experiment, and the formulation, testing, and modification of hypotheses". Experiments need to be designed to test hypotheses. The most important part of the scientific method is the experiment.

The scientific method is a continuous process, which usually begins with observations about the natural world. Human beings are naturally inquisitive, so they often come up with questions about things they see or hear and often develop ideas (hypotheses) about why things are the way they are. The best hypotheses lead to predictions that can be tested in various ways, including making further observations about nature. In general, the strongest tests of hypotheses come from carefully controlled and replicated experiments that gather empirical data. Depending on how well the tests match the predictions, the original hypothesis may require refinement, alteration, expansion or even rejection. If a particular hypothesis becomes very well supported a general theory may be developed.

Although procedures vary from one field of inquiry to another, identifiable features are frequently shared in common between them. The overall process of the scientific method involves making conjectures (hypotheses), deriving predictions from them as logical consequences, and then carrying out experiments based on those predictions. A hypothesis is a conjecture, based on knowledge obtained while formulating the question. The hypothesis might be very specific or it might be broad. Scientists then test hypotheses by conducting experiments. Under modern interpretations, a scientific hypothesis must be falsifiable, implying that it is possible to identify a possible outcome of an experiment that conflicts with predictions deduced from the hypothesis; otherwise, the hypothesis cannot be meaningfully tested.

The purpose of an experiment is to determine whether observations agree with or conflict with the predictions derived from a hypothesis. Experiments can take place anywhere from a college lab to CERN's Large Hadron Collider. There are difficulties in a formulaic statement of method, however. Though the scientific method is often presented as a fixed sequence of steps, it represents rather a set of general principles. Not all steps take place in every scientific inquiry (or to the same degree), and are not always in the same order. Some philosophers and scientists have argued that there is no scientific method, such as Lee Smolin and Paul Feyerabend (in his *Against Method*). Nola and Sankey remark that "For some, the whole idea of a theory of scientific method is yester-year's debate".

Overview

The DNA example below is a synopsis of this method

Ibn al-Haytham (Alhazen), 965–1039 Iraq. A polymath, considered by some to be the father of modern scientific methodology, due to his emphasis on experimental data and reproducibility of its results.

Johannes Kepler (1571–1630). "Kepler shows his keen logical sense in detailing the whole process by which he finally arrived at the true orbit. This is the greatest piece of Retroductive reasoning ever performed." – C. S. Peirce, c. 1896, on Kepler's reasoning through explanatory hypotheses

According to Morris Kline, "Modern science owes its present flourishing state to a new scientific method which was fashioned almost entirely by Galileo Galilei" (1564–1642). Dudley Shapere takes a more measured view of Galileo's contribution.

The scientific method is the process by which science is carried out. As in other areas of inquiry, science (through the scientific method) can build on previous knowledge and develop a more sophisticated understanding of its topics of study over time. This model can be seen to underlay the scientific revolution. One thousand years ago, Alhazen argued the importance of forming questions and subsequently testing them, an approach which was advocated by Galileo in 1638 with the publication of *Two New Sciences*. The current method is based on a hypothetico-deductive model formulated in the 20th century, although it has undergone significant revision since first proposed (for a more formal discussion).

Process

The overall process involves making conjectures (hypotheses), deriving predictions from them as logical consequences, and then carrying out experiments based on those predictions to determine whether the original conjecture was correct. There are difficulties in a formulaic statement of method, however. Though the scientific method is often presented as a fixed sequence of steps, they are better considered as general principles. Not all steps take place in every scientific inquiry (or to the same degree), and are not always in the same order. As noted by William Whewell (1794–1866), "invention, sagacity, [and] genius" are required at every step.

Formulation of a Question

The question can refer to the explanation of a specific *observation*, as in "Why is the sky blue?", but can also be open-ended, as in "How can I design a drug to cure this particular disease?" This stage frequently involves finding and evaluating evidence from previous experiments, personal scientific observations or assertions, and/or the work of other scientists. If the answer is already known, a different question that builds on the previous evidence can be posed. When applying the scientific method to scientific research, determining a good question can be very difficult and affects the final outcome of the investigation.

Hypothesis

A hypothesis is a conjecture, based on knowledge obtained while formulating the question, that may explain the observed behavior of a part of our universe. The hypothesis might be very specific, e.g., Einstein's equivalence principle or Francis Crick's "DNA makes RNA makes protein", or it might be broad, e.g., unknown species of life dwell in the unexplored depths of the oceans. A statistical hypothesis is a conjecture about some population. For example, the population might be people with a particular disease. The conjecture might be that a new drug will cure the disease in some of those people. Terms commonly associated with statistical hypotheses are null hypothesis and alternative hypothesis. A null hypothesis is the conjecture that the statistical hypothesis is false, e.g., that the new drug does nothing and that any cures are due to chance effects. Researchers normally want to show that the null hypothesis is false. The alternative hypothesis is the desired outcome, e.g., that the drug does better than chance. A final point: a scientific hypothesis must be falsifiable, meaning that one can identify a possible outcome of an experiment that conflicts with predictions deduced from the hypothesis; otherwise, it cannot be meaningfully tested.

Prediction

This step involves determining the logical consequences of the hypothesis. One or more predictions are then selected for further testing. The more unlikely that a prediction would be correct simply by coincidence, then the more convincing it would be if the prediction were fulfilled; evidence is also stronger if the answer to the prediction is not already known, due to the effects of hindsight bias. Ideally, the prediction must also distinguish the hypothesis from likely alternatives; if two hypotheses make the same prediction, observing the prediction to be correct is not evidence for either one over the other. (These statements about the relative strength of evidence can be mathematically derived using Bayes' Theorem).

Testing

This is an investigation of whether the real world behaves as predicted by the hypothesis. Scientists (and other people) test hypotheses by conducting experiments. The purpose of an experiment is to determine whether observations of the real world agree with or conflict with the predictions derived from a hypothesis. If they agree, confidence in the hypothesis increases; otherwise, it decreases. Agreement does not assure that the hypothesis is true; future experiments may reveal problems. Karl Popper advised scientists to try to falsify hypotheses, i.e., to search for and test those experiments that seem most doubtful. Large numbers of successful confirmations are not convincing if they arise from experiments that avoid risk. Experiments should be designed to minimize possible errors, especially through the use of appropriate scientific controls. For example, tests of medical treatments are commonly run as double-blind tests. Test personnel, who might unwittingly reveal to test subjects which samples are the desired test drugs and which are placebos, are kept ignorant of which are which. Such hints can bias the responses of the test subjects. Furthermore, failure of an experiment does not necessarily mean the hypothesis is false. Experiments always depend on several hypotheses, e.g., that the test equipment is working properly, and a failure may be a failure of one of the auxiliary hypotheses. Experiments can be conducted in a college lab, on a kitchen table, at CERN's Large Hadron Collider, at the bottom of an ocean, on Mars (using one of the working rovers), and so on. Astronomers do experiments, searching for planets around distant stars. Finally, most individual experiments address highly specific topics for reasons of practicality. As a result, evidence about broader topics is usually accumulated gradually.

Analysis

This involves determining what the results of the experiment show and deciding on the next actions to take. The predictions of the hypothesis are compared to those of the null hypothesis, to determine which is better able to explain the data. In cases where an experiment is repeated many times, a statistical analysis such as a chi-squared test may be required. If the evidence has falsified the hypothesis, a new hypothesis is required; if the experiment supports the hypothesis but the evidence is not strong enough for high confidence, other predictions from the hypothesis must be tested. Once a hypothesis is strongly supported by evidence, a new question can be asked to provide further insight on the same topic. Evidence from other scientists and experience are frequently incorporated at any stage in the process. Depending on the complexity of the experiment, many iterations may be required to gather sufficient evidence to answer a question with confidence, or to build up many answers to highly specific questions in order to answer a single broader question.

DNA Example

The discovery became the starting point for many further studies involving the genetic material, such as the field of molecular genetics, and it was awarded the Nobel Prize in 1962. Each step of the example is examined in more detail later in the article.

Other Components

The scientific method also includes other components required even when all the iterations of the steps above have been completed:

Replication

If an experiment cannot be repeated to produce the same results, this implies that the original results might have been in error. As a result, it is common for a single experiment to be performed multiple times, especially when there are uncontrolled variables or other indications of experimental error. For significant or surprising results, other scientists may also attempt to replicate the results for themselves, especially if those results would be important to their own work.

External Review

The process of peer review involves evaluation of the experiment by experts, who typically give their opinions anonymously. Some journals request that the experimenter provide lists of possible peer reviewers, especially if the field is highly specialized. Peer review does not certify correctness of the results, only that, in the opinion of the reviewer, the experiments themselves were sound (based on the description supplied by the experimenter). If the work passes peer review, which occasionally may require new experiments requested by the reviewers, it will be published in a peer-reviewed scientific journal. The specific journal that publishes the results indicates the perceived quality of the work.

Data Recording and Sharing

Scientists typically are careful in recording their data, a requirement promoted by Ludwik Fleck (1896–1961) and others. Though not typically required, they might be requested to supply this data to other scientists who wish to replicate their original results (or parts of their original results), extending to the sharing of any experimental samples that may be difficult to obtain.

Scientific Inquiry

Scientific inquiry generally aims to obtain knowledge in the form of testable explanations that scientists can use to predict the results of future experiments. This allows scientists to gain a better understanding of the topic under study, and later to use that understanding to intervene in its causal mechanisms (such as to cure disease). The better an explanation is at making predictions, the more useful it frequently can be, and the more likely it will continue to explain a body of evidence better than its alternatives. The most successful explanations - those which explain and make accurate predictions in a wide range of circumstances - are often called scientific theories.

Most experimental results do not produce large changes in human understanding; improvements in theoretical scientific understanding typically result from a gradual process of development over time, sometimes across different domains of science. Scientific models vary in the extent to which they have been experimentally tested and for how long, and in their acceptance in the scientific community. In general, explanations become accepted over time as evidence accumulates on a given topic, and the explanation in question proves more powerful than its alternatives at explaining the evidence.

Often subsequent researchers re-formulate the explanations over time, or combined explanations to produce new explanations.

Tow sees the scientific method in terms of an evolutionary algorithm applied to science and technology.

Properties of Scientific Inquiry

Scientific knowledge is closely tied to empirical findings, and can remain subject to falsification if new experimental observation incompatible with it is found. That is, no theory can ever be considered final, since new problematic evidence might be discovered. If such evidence is found, a new theory may be proposed, or (more commonly) it is found that modifications to the previous theory are sufficient to explain the new evidence. The strength of a theory can be argued to relate to how long it has persisted without major alteration to its core principles.

Theories can also become subsumed by other theories. For example, Newton's laws explained thousands of years of scientific observations of the planets almost perfectly. However, these laws were then determined to be special cases of a more general theory (relativity), which explained both the (previously unexplained) exceptions to Newton's laws and predicted and explained other observations such as the deflection of light by gravity. Thus, in certain cases independent, unconnected, scientific observations can be connected to each other, unified by principles of increasing explanatory power.

Since new theories might be more comprehensive than what preceded them, and thus be able to explain more than previous ones, successor theories might be able to meet a higher standard by explaining a larger body of observations than their predecessors. For example, the theory of evolution explains the diversity of life on Earth, how species adapt to their environments, and many other patterns observed in the natural world; its most recent major modification was unification with genetics to form the modern evolutionary synthesis. In subsequent modifications, it has also subsumed aspects of many other fields such as biochemistry and molecular biology.

Beliefs and Biases

Flying gallop falsified

Scientific Method

Muybridge's photographs of *The Horse in Motion,* 1878, were used to answer the question whether all four feet of a galloping horse are ever off the ground at the same time. This demonstrates a use of photography in science.

Scientific methodology often directs that hypotheses be tested in controlled conditions wherever possible. This is frequently possible in certain areas, such as in the biological sciences, and more difficult in other areas, such as in astronomy. The practice of experimental control and reproducibility can have the effect of diminishing the potentially harmful effects of circumstance, and to a degree, personal bias. For example, pre-existing beliefs can alter the interpretation of results, as in confirmation bias; this is a heuristic that leads a person with a particular belief to see things as reinforcing their belief, even if another observer might disagree (in other words, people tend to observe what they expect to observe).

A historical example is the belief that the legs of a galloping horse are splayed at the point when none of the horse's legs touches the ground, to the point of this image being included in paintings by its supporters. However, the first stop-action pictures of a horse's gallop by Eadweard Muybridge showed this to be false, and that the legs are instead gathered together. Another important human bias that plays a role is a preference for new, surprising statements, which can result in a search for evidence that the new is true. In contrast to this standard in the scientific method, poorly attested beliefs can be believed and acted upon via a less rigorous heuristic, sometimes taking advantage of the narrative fallacy that when narrative is constructed its elements become easier to believe. Sometimes, these have their elements assumed *a priori*, or contain some other logical or methodological flaw in the process that ultimately produced them.

Elements of the Scientific Method

There are different ways of outlining the basic method used for scientific inquiry. The scientific community and philosophers of science generally agree on the following classification of method components. These methodological elements and organization of procedures tend to be more characteristic of natural sciences than social sciences. Nonetheless, the cycle of formulating hypotheses, testing and analyzing the results, and formulating new hypotheses, will resemble the cycle described below.

Four essential elements of the scientific method are iterations, recursions, interleavings, or orderings of the following:

- Characterizations (observations, definitions, and measurements of the subject of inquiry)
- Hypotheses (theoretical, hypothetical explanations of observations and measurements of the subject)
- Predictions (reasoning including deductive reasoning from the hypothesis or theory)
- Experiments (tests of all of the above)

Each element of the scientific method is subject to peer review for possible mistakes. These activities do not describe all that scientists do but apply mostly to experimental sciences (e.g., physics, chemistry, and biology). The elements above are often taught in the educational system as "the scientific method".

The scientific method is not a single recipe: it requires intelligence, imagination, and creativity. In this sense, it is not a mindless set of standards and procedures to follow, but is rather an ongoing cycle, constantly developing more useful, accurate and comprehensive models and methods. For example, when Einstein developed the Special and General Theories of Relativity, he did not in any way refute or discount Newton's *Principia*. On the contrary, if the astronomically large, the vanishingly small, and the extremely fast are removed from Einstein's theories – all phenomena Newton could not have observed – Newton's equations are what remain. Einstein's theories are expansions and refinements of Newton's theories and, thus, increase our confidence in Newton's work.

A linearized, pragmatic scheme of the four points above is sometimes offered as a guideline for proceeding:

1. Define a question
2. Gather information and resources (observe)
3. Form an explanatory hypothesis
4. Test the hypothesis by performing an experiment and collecting data in a reproducible manner
5. Analyze the data
6. Interpret the data and draw conclusions that serve as a starting point for new hypothesis
7. Publish results

8. Retest (frequently done by other scientists)

The iterative cycle inherent in this step-by-step method goes from point 3 to 6 back to 3 again.

While this schema outlines a typical hypothesis/testing method, it should also be noted that a number of philosophers, historians and sociologists of science (perhaps most notably Paul Feyerabend) claim that such descriptions of scientific method have little relation to the ways that science is actually practiced.

Characterizations

The scientific method depends upon increasingly sophisticated characterizations of the subjects of investigation. (The *subjects* can also be called *unsolved problems* or the *unknowns*.) For example, Benjamin Franklin conjectured, correctly, that St. Elmo's fire was electrical in nature, but it has taken a long series of experiments and theoretical changes to establish this. While seeking the pertinent properties of the subjects, careful thought may also entail some definitions and observations; the observations often demand careful measurements and/or counting.

The systematic, careful collection of measurements or counts of relevant quantities is often the critical difference between pseudo-sciences, such as alchemy, and science, such as chemistry or biology. Scientific measurements are usually tabulated, graphed, or mapped, and statistical manipulations, such as correlation and regression, performed on them. The measurements might be made in a controlled setting, such as a laboratory, or made on more or less inaccessible or unmanipulatable objects such as stars or human populations. The measurements often require specialized scientific instruments such as thermometers, spectroscopes, particle accelerators, or voltmeters, and the progress of a scientific field is usually intimately tied to their invention and improvement.

I am not accustomed to saying anything with certainty after only one or two observations.

—Andreas Vesalius, (1546)

Uncertainty

Measurements in scientific work are also usually accompanied by estimates of their uncertainty. The uncertainty is often estimated by making repeated measurements of the desired quantity. Uncertainties may also be calculated by consideration of the uncertainties of the individual underlying quantities used. Counts of things, such as the number of people in a nation at a particular time, may also have an uncertainty due to data collection limitations. Or counts may represent a sample of desired quantities, with an uncertainty that depends upon the sampling method used and the number of samples taken.

Definition

Measurements demand the use of *operational definitions* of relevant quantities. That is, a scientific quantity is described or defined by how it is measured, as opposed to some more vague, inexact or "idealized" definition. For example, electric current, measured in amperes, may be operationally defined in terms of the mass of silver deposited in a certain time on an electrode in an electrochemical device that is described in some detail. The operational definition of a thing often relies on comparisons with standards: the operational definition of "mass" ultimately relies on the use of an artifact, such as a particular kilogram of platinum-iridium kept in a laboratory in France.

The scientific definition of a term sometimes differs substantially from its natural language usage. For example, mass and weight overlap in meaning in common discourse, but have distinct meanings in mechanics. Scientific quantities are often characterized by their units of measure which can later be described in terms of conventional physical units when communicating the work.

New theories are sometimes developed after realizing certain terms have not previously been sufficiently clearly defined. For example, Albert Einstein's first paper on relativity begins by defining simultaneity and the means for determining length. These ideas were skipped over by Isaac Newton with, "I do not define time, space, place and motion, as being well known to all." Einstein's paper then demonstrates that they (viz., absolute time and length independent of motion) were approximations. Francis Crick cautions us that when characterizing a subject, however, it can be premature to define something when it remains ill-understood. In Crick's study of consciousness, he actually found it easier to study awareness in the visual system, rather than to study free will, for example. His cautionary example was the gene; the gene was much more poorly understood before Watson and Crick's pioneering discovery of the structure of DNA; it would have been counterproductive to spend much time on the definition of the gene, before them.

DNA-characterization

The history of the discovery of the structure of DNA is a classic example of the elements of the scientific method: in 1950 it was known that genetic inheritance had a mathematical description, starting with the studies of Gregor Mendel, and that DNA contained genetic information (Oswald Avery's *transforming principle*). But the mechanism of storing genetic information (i.e., genes) in DNA was unclear. Researchers in Bragg's laboratory at Cambridge University made X-ray diffraction pictures of various molecules, starting with crystals of salt, and proceeding to more complicated substances. Using clues painstakingly assembled over decades, beginning with its chemical composition, it was determined that it should be possible to characterize the physical structure of DNA, and the X-ray images would be the vehicle. *..2. DNA-hypotheses*

Another Example: Precession of Mercury

Scientific Method

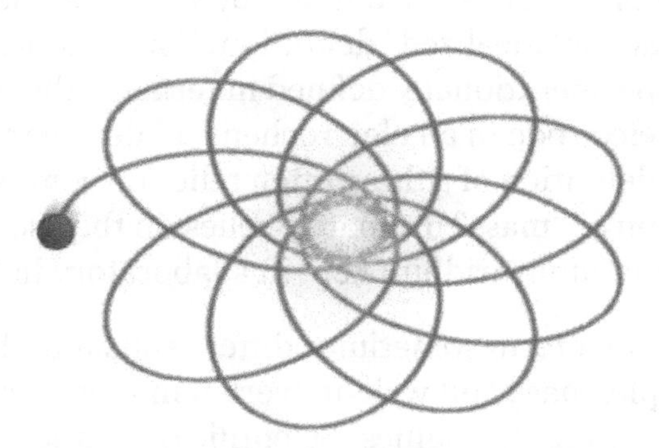

Precession of the perihelion (exaggerated)

The characterization element can require extended and extensive study, even centuries. It took thousands of years of measurements, from the Chaldean, Indian, Persian, Greek, Arabic and European astronomers, to fully record the motion of planet Earth. Newton was able to include those measurements into consequences of his laws of motion. But the perihelion of the planet Mercury's orbit exhibits a precession that cannot be fully explained by Newton's laws of motion, as Leverrier pointed out in 1859. The observed difference for Mercury's precession between Newtonian theory and observation was one of the things that occurred to Einstein as a possible early test of his theory of General Relativity. His relativistic calculations matched observation much more closely than did Newtonian theory. The difference is approximately 43 arc-seconds per century.

Hypothesis Development

A hypothesis is a suggested explanation of a phenomenon, or alternately a reasoned proposal suggesting a possible correlation between or among a set of phenomena.

Normally hypotheses have the form of a mathematical model. Sometimes, but not always, they can also be formulated as existential statements, stating that some particular instance of the phenomenon being studied has some characteristic and causal explanations, which have the general form of universal statements, stating that every instance of the phenomenon has a particular characteristic.

Scientists are free to use whatever resources they have – their own creativity, ideas from other fields, inductive reasoning, Bayesian inference, and so on – to imagine possible explanations for a phenomenon under study. Charles Sanders Peirce, borrowing a page from Aristotle (*Prior Analytics*, 2.25) described the incipient stages of inquiry, instigated by the "irritation of doubt" to venture a plausible guess, as *abductive reasoning*. The history of science is filled with stories of scientists claiming a "flash of inspiration", or a hunch, which then motivated them to look for evidence to support or refute their idea. Michael Polanyi made such creativity the centerpiece of his discussion of methodology.

William Glen observes that

> the success of a hypothesis, or its service to science, lies not simply in its perceived "truth", or power to displace, subsume or reduce a predecessor idea, but perhaps more in its ability to stimulate the research that will illuminate ... bald suppositions and areas of vagueness.

In general scientists tend to look for theories that are "elegant" or "beautiful". In contrast to the usual English use of these terms, they here refer to a theory in accordance with the known facts, which is nevertheless relatively simple and easy to handle. Occam's Razor serves as a rule of thumb for choosing the most desirable amongst a group of equally explanatory hypotheses.

DNA-hypotheses

Linus Pauling proposed that DNA might be a triple helix. This hypothesis was also considered by Francis Crick and James D. Watson but discarded. When Watson and Crick learned of Pauling's hypothesis, they understood from existing data that Pauling was wrong and that Pauling would soon admit his difficulties with that structure. So, the race was on to figure out the correct structure (except that Pauling did not realize at the time that he was in a race) *..3. DNA-predictions*

Predictions from the Hypothesis

Any useful hypothesis will enable predictions, by reasoning including deductive reasoning. It might predict the outcome of an experiment in a laboratory setting or the observation of a phenomenon in nature. The prediction can also be statistical and deal only with probabilities.

It is essential that the outcome of testing such a prediction be currently unknown. Only in this case does a successful outcome increase the probability that the hypothesis is true. If the outcome is already known, it is called a consequence and should have already been considered while formulating the hypothesis.

If the predictions are not accessible by observation or experience, the hypothesis is not yet testable and so will remain to that extent unscientific in a strict sense. A new technology or theory might make the necessary experiments feasible. Thus, much scientifically based speculation might convince one (or many) that the hypothesis that other intelligent species exist is true. But since there no experiment now known which can test this hypothesis, science itself can have little to say about the possibility. In future, some new technique might lead to an experimental test and the speculation would then become part of accepted science.

DNA-predictions

James D. Watson, Francis Crick, and others hypothesized that DNA had a helical structure. This implied that DNA's X-ray diffraction pattern would be 'x shaped'. This prediction followed from the work of Cochran, Crick and Vand (and independently by Stokes). The Cochran-Crick-Vand-Stokes theorem provided a mathematical explanation for the empirical observation that diffraction from helical structures produces x shaped patterns.

In their first paper, Watson and Crick also noted that the double helix structure they proposed provided a simple mechanism for DNA replication, writing, "It has not escaped our notice that the specific pairing we have postulated immediately suggests a possible copying mechanism for the genetic material". ..*4. DNA-experiments*

Another example: General Relativity

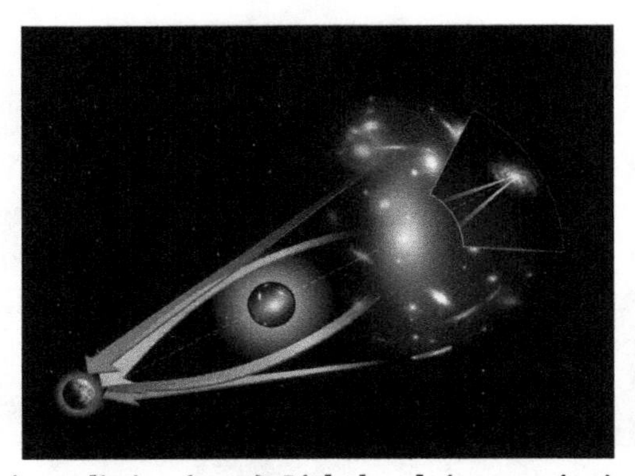

Einstein's prediction (1907): Light bends in a gravitational field

Einstein's theory of General Relativity makes several specific predictions about the observable structure of space-time, such as that light bends in a gravitational field, and that the amount of bending depends in a precise way on the strength of that gravitational field. Arthur Eddington's observations made during a 1919 solar eclipse supported General Relativity rather than Newtonian gravitation.

Experiments

Once predictions are made, they can be sought by experiments. If the test results contradict the predictions, the hypotheses which entailed them are called into question and become less tenable. Sometimes the experiments are conducted incorrectly or are not very well designed, when compared to a crucial experiment. If the experimental results confirm the predictions, then the hypotheses are considered more likely to be correct, but might still be wrong and continue to be subject to further testing. The experimental control is a technique for dealing with observational error. This technique uses the contrast between multiple samples (or observations) under differing conditions to see what varies or what remains the same. We vary the conditions for each measurement, to help isolate what has changed. Mill's canons can then help us figure out what the important factor is. Factor analysis is one technique for discovering the important factor in an effect.

Depending on the predictions, the experiments can have different shapes. It could be a classical experiment in a laboratory setting, a double-blind study or an archaeological excavation. Even taking a plane from New York to Paris is an experiment which tests the aerodynamical hypotheses used for constructing the plane.

Scientists assume an attitude of openness and accountability on the part of those conducting an experiment. Detailed record keeping is essential, to aid in recording and reporting on the experimental results, and supports the effectiveness and integrity of the procedure. They will also assist in reproducing the experimental results, likely by others. Traces of this approach can be seen in the work of Hipparchus (190–120 BCE), when determining a value for the precession of the Earth, while controlled experiments can be seen in the works of Jābir ibn Hayyān (721–815 CE), al-Battani (853–929) and Alhazen (965–1039).

DNA-experiments

Watson and Crick showed an initial (and incorrect) proposal for the structure of DNA to a team from Kings College – Rosalind Franklin, Maurice Wilkins, and Raymond Gosling. Franklin immediately spotted the flaws which concerned the water content. Later Watson saw Franklin's detailed X-ray diffraction images which showed an X-shape and was able to confirm the structure was helical. This rekindled Watson and Crick's model building and led to the correct structure. *..1. DNA-characterizations*

Evaluation and Improvement

The scientific method is iterative. At any stage it is possible to refine its accuracy and precision, so that some consideration will lead the scientist to repeat an earlier part of the process. Failure to develop an interesting hypothesis may lead a scientist to re-define the subject under consideration. Failure of a hypothesis to produce interesting and testable predictions may lead to reconsideration of the hypothesis or of the definition of the subject. Failure of an experiment to produce interesting results may lead a scientist to reconsider the experimental method, the hypothesis, or the definition of the subject.

Other scientists may start their own research and enter the process at any stage. They might adopt the characterization and formulate their own hypothesis, or they might adopt the hypothesis and deduce their own predictions. Often the experiment is not done by the person who made the prediction, and the characterization is based on experiments done by someone else. Published results of experiments can also serve as a hypothesis predicting their own reproducibility.

DNA-iterations

After considerable fruitless experimentation, being discouraged by their superior from continuing, and numerous false starts, Watson and Crick were able to infer the essential structure of DNA by concrete modeling of the physical shapes of the nucleotides which comprise it. They were guided by the bond lengths which had been deduced by Linus Pauling and by Rosalind Franklin's X-ray diffraction images. ..*DNA Example*

Confirmation

Science is a social enterprise, and scientific work tends to be accepted by the scientific community when it has been confirmed. Crucially, experimental and theoretical results must be reproduced by others within the scientific community. Researchers have given their lives for this vision; Georg Wilhelm Richmann was killed by ball lightning (1753) when attempting to replicate the 1752 kite-flying experiment of Benjamin Franklin.

To protect against bad science and fraudulent data, government research-granting agencies such as the National Science Foundation, and science journals, including *Nature* and *Science*, have a policy that researchers must archive their data and methods so that other researchers can test the data and methods and build on the research that has gone before. Scientific data archiving can be done at a number of national archives in the U.S. or in the World Data Center.

Models of Scientific Inquiry

Classical Model

The classical model of scientific inquiry derives from Aristotle, who distinguished the forms of approximate and exact reasoning, set out the threefold scheme of abductive, deductive, and inductive inference, and also treated the compound forms such as reasoning by analogy.

Pragmatic Model

In 1877, Charles Sanders Peirce characterized inquiry in general not as the pursuit of truth *per se* but as the struggle to move from irritating, inhibitory doubts born of surprises, disagreements, and the like, and to reach a secure belief, belief being that on which one is prepared to act. He framed scientific inquiry as part of a broader spectrum and as spurred, like inquiry generally, by actual doubt, not mere verbal or hyperbolic doubt, which he held to be fruitless. He outlined four methods of settling opinion, ordered from least to most successful:

1. The method of tenacity (policy of sticking to initial belief) – which brings comforts and decisiveness but leads to trying to ignore contrary information and others' views as if truth were intrinsically private, not public. It goes against

the social impulse and easily falters since one may well notice when another's opinion is as good as one's own initial opinion. Its successes can shine but tend to be transitory.

2. The method of authority – which overcomes disagreements but sometimes brutally. Its successes can be majestic and long-lived, but it cannot operate thoroughly enough to suppress doubts indefinitely, especially when people learn of other societies present and past.

3. The method of the *a priori* – which promotes conformity less brutally but fosters opinions as something like tastes, arising in conversation and comparisons of perspectives in terms of "what is agreeable to reason." Thereby it depends on fashion in paradigms and goes in circles over time. It is more intellectual and respectable but, like the first two methods, sustains accidental and capricious beliefs, destining some minds to doubt it.

4. The scientific method – the method wherein inquiry regards itself as fallible and purposely tests itself and criticizes, corrects, and improves itself.

Peirce held that slow, stumbling ratiocination can be dangerously inferior to instinct and traditional sentiment in practical matters, and that the scientific method is best suited to theoretical research, which in turn should not be trammeled by the other methods and practical ends; reason's "first rule" is that, in order to learn, one must desire to learn and, as a corollary, must not block the way of inquiry. The scientific method excels the others by being deliberately designed to arrive – eventually – at the most secure beliefs, upon which the most successful practices can be based. Starting from the idea that people seek not truth *per se* but instead to subdue irritating, inhibitory doubt, Peirce showed how, through the struggle, some can come to submit to truth for the sake of belief's integrity, seek as truth the guidance of potential practice correctly to its given goal, and wed themselves to the scientific method.

For Peirce, rational inquiry implies presuppositions about truth and the real; to reason is to presuppose (and at least to hope), as a principle of the reasoner's self-regulation, that the real is discoverable and independent of our vagaries of opinion. In that vein he defined truth as the correspondence of a sign (in particular, a proposition) to its object and, pragmatically, not as actual consensus of some definite, finite community (such that to inquire would be to poll the experts), but instead as that final opinion which all investigators *would* reach sooner or later but still inevitably, if they were to push investigation far enough, even when they start from different points. In tandem he defined the real as a true sign's object (be that object a possibility or quality, or an actuality or brute fact, or a necessity or norm or law), which is what it is independently of any finite community's opinion and, pragmatically, depends only on the final opinion destined in a sufficient investigation. That is a destination as far, or near, as the truth itself to you or me or the given finite community. Thus, his theory of inquiry boils down to "Do the science." Those conceptions of truth and the real involve the idea of a community both without definite limits (and thus potentially self-correcting as far as needed) and capable of definite increase of knowledge. As inference, "logic is rooted in the social principle" since it depends on a standpoint that is, in a sense, unlimited.

Paying special attention to the generation of explanations, Peirce outlined the scientific method as a coordination of three kinds of inference in a purposeful cycle aimed at settling doubts, as follows (in §III–IV in "A Neglected Argument" except as otherwise noted):

1. Abduction (or retroduction). Guessing, inference to explanatory hypotheses for selection of those best worth trying. From abduction, Peirce distinguishes induction as inferring, on the basis of tests, the proportion of truth in the hypothesis. Every inquiry, whether into ideas, brute facts, or norms and laws, arises from surprising observations in one or more of those realms (and for example at any stage of an inquiry already underway). All explanatory content of theories comes from abduction, which guesses a new or outside idea so as to account in a simple, economical way for a surprising or complicative phenomenon. Oftenest, even a well-prepared mind guesses wrong. But the modicum of success of our guesses far exceeds that of sheer luck and seems born of attunement to nature by instincts developed or inherent, especially insofar as best guesses are optimally plausible and simple in the sense, said Peirce, of the "facile and natural", as by Galileo's natural light of reason and as distinct from "logical simplicity". Abduction is the most fertile but least secure mode of inference. Its general rationale is inductive: it succeeds often enough and, without it, there is no hope of sufficiently expediting inquiry (often multi-generational) toward new truths. Coordinative method leads from abducing a plausible hypothesis to judging it for its testability and for how its trial would economize inquiry itself. Peirce calls his pragmatism "the logic of abduction". His pragmatic maxim is: "Consider what effects that might conceivably have practical bearings you conceive the objects of your conception to have. Then, your conception of those effects is the whole of your conception of the object". His pragmatism is a method of reducing conceptual confusions fruitfully by equating the meaning of any conception with the conceivable practical implications of its object's conceived effects—a method of experimentational mental reflection hospitable to forming hypotheses and conducive to testing them. It favors efficiency. The hypothesis, being insecure, needs to have practical implications leading at least to mental tests and, in science, lending themselves to scientific tests. A simple but unlikely guess, if uncostly to test for falsity, may belong first in line for testing. A guess is intrinsically worth testing if it has instinctive plausibility or reasoned objective probability, while subjective likelihood, though reasoned, can be misleadingly seductive. Guesses can be chosen for trial strategically, for their caution (for which Peirce gave as example the game of Twenty Questions), breadth, and incomplexity. One can hope to discover only that which

time would reveal through a learner's sufficient experience anyway, so the point is to expedite it; the economy of research is what demands the leap, so to speak, of abduction and governs its art.

2. Deduction. Two stages:

   i. Explication. Unclearly premissed, but deductive, analysis of the hypothesis in order to render its parts as clear as possible.

   ii. Demonstration: Deductive Argumentation, Euclidean in procedure. Explicit deduction of hypothesis's consequences as predictions, for induction to test, about evidence to be found. Corollarial or, if needed, Theorematic.

3. Induction. The long-run validity of the rule of induction is deducible from the principle (presuppositional to reasoning in general) that the real is only the object of the final opinion to which adequate investigation would lead; anything to which no such process would ever lead would not be real. Induction involving ongoing tests or observations follows a method which, sufficiently persisted in, will diminish its error below any predesignate degree. Three stages:

   i. Classification. Unclearly premissed, but inductive, classing of objects of experience under general ideas.

   ii. Probation: direct inductive argumentation. Crude (the enumeration of instances) or gradual (new estimate of proportion of truth in the hypothesis after each test). Gradual induction is qualitative or quantitative; if qualitative, then dependent on weightings of qualities or characters; if quantitative, then dependent on measurements, or on statistics, or on countings.

   iii. Sentential Induction. "...which, by inductive reasonings, appraises the different probations singly, then their combinations, then makes self-appraisal of these very appraisals themselves, and passes final judgment on the whole result".

Communication and Community

Frequently the scientific method is employed not only by a single person, but also by several people cooperating directly or indirectly. Such cooperation can be regarded as an important element of a scientific community. Various standards of scientific methodology are used within such an environment.

Peer Review Evaluation

Scientific journals use a process of *peer review*, in which scientists' manuscripts are submitted by editors of scientific journals to (usually one to three) fellow (usually anonymous) scientists familiar with the field for evaluation. In certain journals, the journal itself selects the referees; while in others (especially journals that are extremely specialized), the manuscript author might recommend referees. The referees may or may not recommend publication, or they might recommend publication with suggested modifications, or sometimes, publication in another journal. This standard is practiced to various degrees by different journals, and can have the effect of keeping the literature free of obvious errors and to generally improve the quality of the material, especially in the journals who use the standard most rigorously. The peer review process can have limitations when considering research outside the conventional scientific paradigm: problems of "groupthink" can interfere with open and fair deliberation of some new research.

Documentation and Replication

Sometimes experimenters may make systematic errors during their experiments, veer from standard methods and practices (Pathological science) for various reasons, or, in rare cases, deliberately report false results. Occasionally because of this then, other scientists might attempt to repeat the experiments in order to duplicate the results.

Archiving

Researchers sometimes practice scientific data archiving, such as in compliance with the policies of government funding agencies and scientific journals. In these cases, detailed records of their experimental procedures, raw data, statistical analyses and source code can be preserved in order to provide evidence of the methodology and practice of the procedure and assist in any potential future attempts to reproduce the result. These procedural records may also assist in the conception of new experiments to test the hypothesis, and may prove useful to engineers who might examine the potential practical applications of a discovery.

Data Sharing

When additional information is needed before a study can be reproduced, the author of the study might be asked to provide it. They might provide it, or if the author refuses to share data, appeals can be made to the journal editors who published the study or to the institution which funded the research.

Limitations

Since it is impossible for a scientist to record *everything* that took place in an experiment, facts selected for their apparent relevance are reported. This may lead, unavoidably, to problems later if some supposedly irrelevant feature is questioned. For example, Heinrich Hertz did not report the size of the room used to test Maxwell's equations, which later turned out to account for a small deviation in the results. The problem is that parts of the theory itself need to be assumed in order to select and report the experimental conditions. The observations are hence sometimes described as being 'theory-laden'.

Dimensions of Practice

The primary constraints on contemporary science are:

- Publication, i.e. Peer review
- Resources (mostly funding)

It has not always been like this: in the old days of the "gentleman scientist" funding (and to a lesser extent publication) were far weaker constraints.

Both of these constraints indirectly require scientific method – work that violates the constraints will be difficult to publish and difficult to get funded. Journals require submitted papers to conform to "good scientific practice" and to a degree this can be enforced by peer review. Originality, importance and interest are more important – see for example the author guidelines for *Nature*.

Smaldino and McElreath 2016 have noted that our need to reward scientific understanding is being nullified by poor research design and poor data analysis, which is leading to false-positive findings.

Philosophy and Sociology of Science

Philosophy of science looks at the underpinning logic of the scientific method, at what separates science from non-science, and the ethic that is implicit in science. There are basic assumptions, derived from philosophy by at least one prominent scientist, that form the base of the scientific method – namely, that reality is objective and consistent, that humans have the capacity to perceive reality accurately, and that rational explanations exist for elements of the real world. These assumptions from methodological naturalism form a basis on which science may be grounded. Logical Positivist, empiricist, falsificationist, and other theories have criticized these assumptions and given alternative accounts of the logic of science, but each has also itself been criticized. More generally, the scientific method can be recognized as an idealization.

Thomas Kuhn examined the history of science in his *The Structure of Scientific Revolutions*, and found that the actual method used by scientists differed dramatically from the then-espoused method. His observations of science practice are essentially sociological and do not speak to how science is or can be practiced in other times and other cultures.

Norwood Russell Hanson, Imre Lakatos and Thomas Kuhn have done extensive work on the "theory laden" character of observation. Hanson (1958) first coined the term for the idea that all observation is dependent on the conceptual framework of the observer, using the concept of gestalt to show how preconceptions can affect both observation and description. He opens Chapter 1 with a discussion of the Golgi bodies and their initial rejection as an artefact of staining technique, and a discussion of Brahe and Kepler observing the dawn and seeing a "different" sun rise despite the same physiological phenomenon. Kuhn and Feyerabend acknowledge the pioneering significance of his work.

Kuhn (1961) said the scientist generally has a theory in mind before designing and undertaking experiments so as to make empirical observations, and that the "route from theory to measurement can almost never be traveled backward". This implies that the way in which theory is tested is dictated by the nature of the theory itself, which led Kuhn (1961, p. 166) to argue that "once it has been adopted by a profession ... no theory is recognized to be testable by any quantitative tests that it has not already passed".

Paul Feyerabend similarly examined the history of science, and was led to deny that science is genuinely a methodological process. In his book *Against Method* he argues that scientific progress is *not* the result of applying any particular method. In essence, he says that for any specific method or norm of science, one can find a historic episode where violating it has contributed to the progress of science. Thus, if believers in scientific method wish to express a single universally valid rule, Feyerabend jokingly suggests, it should be 'anything goes'. Criticisms such as his led to the strong programme, a radical approach to the sociology of science.

The postmodernist critiques of science have themselves been the subject of intense controversy. This ongoing debate, known as the science wars, is the result of conflicting values and assumptions between the postmodernist and realist camps. Whereas postmodernists assert that scientific knowledge is simply another discourse (note that this term has special meaning in this context) and not representative of any form of fundamental truth, realists in the scientific community maintain that scientific knowledge does reveal real and fundamental truths about reality. Many books have been written by scientists which take on this problem and challenge the assertions of the postmodernists while defending science as a legitimate method of deriving truth.

Role of Chance in Discovery

Somewhere between 33% and 50% of all scientific discoveries are estimated to have been *stumbled upon*, rather than sought out. This may explain why scientists so often express that they were lucky. Louis Pasteur is credited with the famous saying that "Luck favours the prepared mind", but some psychologists have begun to study what it means to be 'prepared for luck' in the scientific context. Research is showing that scientists are taught various heuristics that tend to harness chance and the unexpected. This is what Nassim Nicholas Taleb calls "Anti-fragility"; while some systems of investigation are fragile in the face of human error, human bias, and randomness, the scientific method is more than resistant or tough – it actually benefits from such randomness in many ways (it is anti-fragile). Taleb believes that the more anti-fragile the system, the more it will flourish in the real world.

Psychologist Kevin Dunbar says the process of discovery often starts with researchers finding bugs in their experiments. These unexpected results lead researchers to try to fix what they *think* is an error in their method. Eventually, the researcher decides the error is too persistent and systematic to be a coincidence. The highly controlled, cautious and curious aspects of the scientific method are thus what make it well suited for identifying such persistent systematic errors. At this point, the researcher will begin to think of theoretical explanations for the error, often seeking the help of colleagues across different domains of expertise.

History

Aristotle, 384 BCE – 322 BCE. "As regards his method, Aristotle is recognized as the inventor of scientific method because of his refined analysis of logical implications contained in demonstrative discourse, which goes well beyond natural logic and does not owe anything to the ones who philosophized before him." – Riccardo Pozzo

The development of the scientific method emerges in the history of science itself. Ancient Egyptian documents describe empirical methods in astronomy, mathematics, and medicine. The Greeks made contributions to the scientific method, most notably through Aristotle in his six works of logic collected as the Organon. Aristotle's inductive-deductive method used inductions from observations to infer general principles, deductions from those principles to check against further observations, and more cycles of induction and deduction to continue the advance of knowledge

According to Karl Popper, Parmenides (*fl.* 5th century BCE) had conceived an axiomatic-deductive method. According to David Lindberg, Aristotle (4th century BCE) wrote about the scientific method even if he and his followers did not actually follow what he said. Lindberg also notes that Ptolemy (2nd century CE) and Ibn al-Haytham (11th century CE) are among the early examples of people who carried out scientific experiments. Also, John Losee writes that "the *Physics* and the *Metaphysics* contain discussions of certain aspects of scientific method", of which, he says "Aristotle viewed scientific inquiry as a progression from observations to general principles and back to observations."

Early Christian leaders such as Clement of Alexandria (150–215) and Basil of Caesarea (330–379) encouraged future generations to view the Greek wisdom as "handmaidens to theology" and science was considered a means to more accurate understanding of the Bible and of God. Augustine of Hippo (354–430) who contributed great philosophical wealth to the Latin Middle Ages, advocated the study of science and was wary of philosophies that disagreed with the Bible, such as astrology and the Greek belief that the world had no beginning. This Christian accommodation with Greek science "laid a foundation for the later widespread, intensive study of natural philosophy during the Late Middle Ages." However, the division of Latin-speaking Western Europe from the Greek-speaking East, followed by barbarian invasions, the Plague of Justinian, and the Islamic conquests, resulted in the West largely losing access to Greek wisdom.

By the 8th century Islam had conquered the Christian lands of Syria, Iraq, Iran and Egypt. This swift conquest further severed Western Europe from many of the great works of Aristotle, Plato, Euclid and others, many of which were housed in the great library of Alexandria. Having come upon such a wealth of knowledge, the Arabs, who viewed non-Arab languages as inferior, even as a source of pollution, employed conquered Christians and Jews to translate these works from the native Greek and Syriac into Arabic.

Thus equipped, Arab philosopher Alhazen (Ibn al-Haytham) performed optical and physiological experiments, reported in his manifold works, the most famous being *Book of Optics* (1021). He was thus a forerunner of scientific method, having understood that a controlled environment involving experimentation and measurement is required in order to draw educated conclusions. Other Arab polymaths of the same era produced copious works on mathematics, philosophy, astronomy and alchemy. Most stuck closely to Aristotle, being hesitant to admit that some of Aristotle's thinking was errant, while others strongly criticized him.

During these years, occasionally a paraphrased translation from the Arabic, which itself had been translated from Greek and Syriac, might make its way to the West for scholarly study. It was not until 1204, during which the Latins conquered and took Constantinople from the Byzantines in the name of the fourth Crusade, that a renewed scholarly interest in the original Greek manuscripts began to grow. Due to the new easier access to the libraries of Constantinople by Western scholars, a certain revival in the study and analysis of the original Greek texts by Western scholars began. From that point a functional scientific method that would launch modern science was on the horizon.

Grosseteste (1175–1253), an English statesman, scientist and Christian theologian, was "the principal figure" in bringing about "a more adequate method of scientific inquiry" by which "medieval scientists were able eventually to outstrip their ancient European and Muslim teachers" (Dales 1973, p. 62). ... His thinking influenced Roger Bacon, who spread Grosseteste's ideas from Oxford to the University of Paris during a visit there in the 1240s. From the prestigious universities in Oxford and Paris, the new experimental science spread rapidly throughout the medieval universities: "And so it went to Galileo, William Gilbert, Francis Bacon, William Harvey, Descartes, Robert Hooke, Newton, Leibniz, and the world of the seventeenth century" (Crombie 1953, p. 15). "So it went to us as well " (Gauch 2003, pp. 52–53).

Roger Bacon (c. 1214 – c. 1292) is sometimes credited as one of the earliest European advocates of the modern scientific method inspired by the works of Aristotle.

Roger Bacon (c. 1214 – c. 1292), an English thinker and experimenter, is recognized by many to be the father of modern scientific method. His view that mathematics was essential to a correct understanding of natural philosophy was considered to be 400 years ahead of its time. He was viewed as "a lone genius proclaiming the truth about time," having correctly calculated the calendar His work in optics provided the platform on which Newton, Descartes, Huygens and others later transformed the science of light. Bacon's groundbreaking advances were due largely to his discovery that experimental science must be based on mathematics. (186–187) His works Opus Majus and De Speculis Comburentibus contain many "carefully drawn diagrams showing Bacon's meticulous investigations into the behavior of light." He gives detailed descriptions of systematic studies using prisms and measurements by which he shows how a rainbow functions.

Others who advanced scientific method during this era included Albertus Magnus (c. 1193 – 1280), Theodoric of Freiberg, (c. 1250 – c. 1310), William of Ockham (c. 1285 – c. 1350), and Jean Buridan (c. 1300 – c. 1358). These were not only scientists but leaders of the church – Christian archbishops, friars and priests.

By the late 15th century, the physician-scholar Niccolò Leoniceno was finding errors in Pliny's *Natural History*. As a physician, Leoniceno was concerned about these botanical errors propagating to the materia medica on which medicines were based. To counter this, a botanical garden was established at Orto botanico di Padova, University of Padua (in use for teaching by 1546), in order that medical students might have empirical access to the plants of a pharmacopia. The philosopher and physician Francisco Sanches was led by his medical training at Rome, 1571–73, and by the philosophical skepticism recently placed in the European mainstream by the publication of Sextus Empiricus' "Outlines of Pyrrhonism", to search for a true method of knowing (*modus sciendi*), as nothing clear can be known by the methods of Aristotle and his followers – for example, syllogism fails upon circular reasoning. Following the physician Galen's *method of medicine*, Sanches lists the methods of judgement and experience, which are faulty in the wrong hands, and we are left with the bleak statement *That Nothing is Known* (1581). This challenge was taken up by René Descartes in the next generation (1637), but at the least, Sanches warns us that we ought to refrain from the methods, summaries, and commentaries on Aristotle, if we seek scientific knowledge. In this, he is echoed by Francis Bacon, also influenced by skepticism; Sanches cites the humanist Juan Luis Vives who sought a better educational system, as well as a statement of human rights as a pathway for improvement of the lot of the poor.

The modern scientific method crystallized no later than in the 17th and 18th centuries. In his work *Novum Organum* (1620) – a reference to Aristotle's *Organon* – Francis Bacon outlined a new system of logic to improve upon the old philosophical process of syllogism. Then, in 1637, René Descartes established the framework for scientific method's guiding principles in his treatise, *Discourse on Method*. The writings of Alhazen, Bacon and Descartes are considered critical in the historical development of the modern scientific method, as are those of John Stuart Mill.

In the late 19th century, Charles Sanders Peirce proposed a schema that would turn out to have considerable influence in the development of current scientific methodology generally. Peirce accelerated the progress on several fronts. Firstly, speaking in broader context in "How to Make Our Ideas Clear" (1878), Peirce outlined an objectively verifiable method to test the truth of putative knowledge on a way that goes beyond mere foundational alternatives, focusing upon both *deduction* and *induction*. He thus placed induction and deduction in a complementary rather than competitive context (the latter of which had been the primary trend at least since David Hume, who wrote in the mid-to-late 18th century). Secondly, and of more direct importance to modern method, Peirce put forth the basic schema for hypothesis/testing that continues to prevail today. Extracting the theory of inquiry from its raw materials in classical logic, he refined it in parallel with the early development of symbolic logic to address the then-current problems in scientific reasoning. Peirce examined and articulated the three fundamental modes of reasoning that, as discussed above in this article, play a role in inquiry today, the processes that are currently known as abductive, deductive, and inductive inference. Thirdly, he played a major role in the progress of symbolic logic itself – indeed this was his primary specialty.

Beginning in the 1930s, Karl Popper argued that there is no such thing as inductive reasoning. All inferences ever made, including in science, are purely deductive according to this view. Accordingly, he claimed that the empirical character of science has nothing to do with induction – but with the deductive property of falsifiability that scientific hypotheses have. Contrasting his views with inductivism and positivism, he even denied the existence of the scientific method: "(1) There is no method of discovering a scientific theory (2) There is no method for ascertaining the truth of a scientific hypothesis, i.e., no method of verification; (3) There is no method for ascertaining whether a hypothesis is 'probable', or probably true". Instead, he held that there is only one universal method, a method not particular to science: The negative method of criticism, or colloquially termed trial and error. It covers not only all products of the human mind, including science, mathematics, philosophy, art and so on, but also the evolution of life. Following Peirce and others, Popper argued that science is fallible and has no authority. In contrast to empiricist-inductivist views, he welcomed metaphysics and philosophical discussion and even gave qualified support to myths and pseudosciences. Popper's view has become known as critical rationalism.

Although science in a broad sense existed before the modern era, and in many historical civilizations (as described above), modern science is so distinct in its approach and successful in its results that it now defines what science is in the strictest sense of the term.

Relationship with Mathematics

Science is the process of gathering, comparing, and evaluating proposed models against observables. A model can be a simulation, mathematical or chemical formula, or set of proposed steps. Science is like mathematics in that researchers in both disciplines can clearly distinguish what is *known* from what is *unknown* at each stage of discovery. Models, in both science and mathematics, need to be internally consistent and also ought to be *falsifiable* (capable of disproof). In mathematics, a statement need not yet be proven; at such a stage, that statement would be called a conjecture. But when a statement has attained mathematical proof, that statement gains a kind of immortality which is highly prized by mathematicians, and for which some mathematicians devote their lives.

Mathematical work and scientific work can inspire each other. For example, the technical concept of time arose in science, and timelessness was a hallmark of a mathematical topic. But today, the Poincaré conjecture has been proven using time as a mathematical concept in which objects can flow.

Nevertheless, the connection between mathematics and reality (and so science to the extent it describes reality) remains obscure. Eugene Wigner's paper, *The Unreasonable Effectiveness of Mathematics in the Natural Sciences*, is a very well known account of the issue from a Nobel Prize-winning physicist. In fact, some observers (including some well known mathematicians such as Gregory Chaitin, and others such as Lakoff and Núñez) have suggested that mathematics is the result of practitioner bias and human limitation (including cultural ones), somewhat like the post-modernist view of science.

George Pólya's work on problem solving, the construction of mathematical proofs, and heuristic show that the mathematical method and the scientific method differ in detail, while nevertheless resembling each other in using iterative or recursive steps.

	Mathematical method	Scientific method
1	Understanding	Characterization from experience and observation
2	Analysis	Hypothesis: a proposed explanation
3	Synthesis	Deduction: prediction from the hypothesis
4	Review/Extend	Test and experiment

In Pólya's view, *understanding* involves restating unfamiliar definitions in your own words, resorting to geometrical figures, and questioning what we know and do not know already; *analysis*, which Pólya takes from Pappus, involves free and heuristic construction of plausible arguments, working backward from the goal, and devising a plan for constructing the proof; *synthesis* is the strict Euclidean exposition of step-by-step details of the proof; *review* involves reconsidering and re-examining the result and the path taken to it.

Gauss, when asked how he came about his theorems, once replied "durch planmässiges Tattonieren" (through systematic palpable experimentation).

Imre Lakatos argued that mathematicians actually use contradiction, criticism and revision as principles for improving their work. In like manner to science, where truth is sought, but certainty is not found, in *Proofs and refutations* (1976), what Lakatos tried to establish was that no theorem of informal mathematics is final or perfect. This means that we should not think that a theorem is ultimately true, only that no counterexample has yet been found. Once a counterexample, i.e. an entity contradicting/not explained by the theorem is found, we adjust the theorem, possibly extending the domain of its validity. This is a continuous way our knowledge accumulates, through the logic and process of proofs and refutations. (If axioms are given for a branch of mathematics, however, Lakatos claimed that proofs from those axioms were tautological, i.e. logically true, by rewriting them, as did Poincaré (*Proofs and Refutations*, 1976).)

Lakatos proposed an account of mathematical knowledge based on Polya's idea of heuristics. In *Proofs and Refutations*, Lakatos gave several basic rules for finding proofs and counterexamples to conjectures. He thought that mathematical 'thought experiments' are a valid way to discover mathematical conjectures and proofs.

Relationship with Statistics

The scientific method has been extremely successful in bringing the world out of medieval thinking, especially once it was combined with industrial processes. However, when the scientific method employs statistics as part of its arsenal, there are mathematical and practical issues that can have a deleterious effect on the reliability of the output of scientific methods. This is described in a popular 2005 scientific paper "Why Most Published Research Findings Are False" by John Ioannidis.

The particular points raised are statistical ("The smaller the studies conducted in a scientific field, the less likely the research findings are to be true" and "The greater the flexibility in designs, definitions, outcomes, and analytical modes in a scientific field, the less likely the research findings are to be true.") and economical ("The greater the financial and other interests and prejudices in a scientific field, the less likely the research findings are to be true" and "The hotter a scientific field (with more scientific teams involved), the less likely the research findings are to be true.") Hence: "Most research findings are false for most research designs and for most fields" and "As shown, the majority of modern biomedical research is operating in areas with very low pre- and poststudy probability for true findings." However: "Nevertheless, most new discoveries will continue to stem from hypothesis-generating research with low or very low pre-study odds," which means that *new* discoveries will come from research that, when that research started, had low or very low odds (a low or very low chance) of succeeding. Hence, if the scientific method is used to expand the frontiers of knowledge, research into areas that are outside the mainstream will yield most new discoveries.

Internal Validity

In scientific research, internal validity is the extent to which a causal conclusion based on a study is warranted, which is determined by the degree to which a study minimizes systematic error (or 'bias'). It contrasts with external validity, the degree to which it is warranted to generalize results to other contexts.

Details

Inferences are said to possess internal validity if a causal relation between two variables is properly demonstrated. A causal inference may be based on a relation when three criteria are satisfied:

1. the "cause" precedes the "effect" in time (temporal precedence),
2. the "cause" and the "effect" are related (covariation), and
3. there are no plausible alternative explanations for the observed covariation (nonspuriousness).

In scientific experimental settings, researchers often manipulate a variable (the independent variable) to see what effect it has on a second variable (the dependent variable). For example, a researcher might, for different experimental groups, manipulate the dosage of a particular drug between groups to see what effect it has on health. In this example, the researcher wants to make a causal inference, namely, that different doses of the drug may be *held responsible* for observed changes or differences. When the researcher may confidently attribute the observed changes or differences in the dependent variable to the independent variable, and when the researcher can rule out other explanations (or *rival hypotheses*), then the causal inference is said to be internally valid.

In many cases, however, the magnitude of effects found in the dependent variable may not just depend on

- variations in the independent variable,
- the power of the instruments and statistical procedures used to measure and detect the effects, and
- the choice of statistical methods.

Rather, a number of variables or circumstances uncontrolled for (or uncontrollable) may lead to additional or alternative explanations (a) for the effects found and/or (b) for the magnitude of the effects found. Internal validity, therefore, is more a matter of degree than of either-or, and that is exactly why research designs other than true experiments may also yield results with a high degree of internal validity.

In order to allow for inferences with a high degree of internal validity, precautions may be taken during the design of the scientific study. As a rule of thumb, conclusions based on correlations or associations may only allow for lesser degrees of internal validity than conclusions drawn on the basis of direct manipulation of the independent variable. And, when viewed only from the perspective of Internal Validity, highly controlled true experimental designs (i.e. with random selection, random assignment to either the control or experimental groups, reliable instruments, reliable manipulation processes, and safeguards against confounding factors) may be the "gold standard" of scientific research. By contrast, however, the very strategies employed to control these factors may also limit the generalizability or External Validity of the findings.

Factors Affecting Internal Validity

- History effect: Events that occur besides the treatment (events in the environment)
- Maturation: Physical or psychological changes in the participants

- Testing: Effect of experience with the pretest - - become test wise.
- Instrumentation: Learning gain might be observed from pre to posttest simply due to nature of the instrument.
- Selection: Effect of treatment confounded with other factors because of selection of participants, problem in non random sample
- Statistical regression: Tendency for participants whose scores fall at either extreme on a variable to score nearer the mean when measured a second time.
- Mortality: Participants lost from the study, attrition.

Threats to Internal Validity

Ambiguous Temporal Precedence

Lack of clarity about which variable occurred first may yield confusion about which variable is the cause and which is the effect.

Confounding

A major threat to the validity of causal inferences is confounding: Changes in the dependent variable may rather be attributed to the existence or variations in the degree of a third variable which is related to the manipulated variable. Where spurious relationships cannot be ruled out, rival hypotheses to the original causal inference hypothesis of the researcher may be developed.

Selection Bias

Selection bias refers to the problem that, at pre-test, differences between groups exist that may interact with the independent variable and thus be 'responsible' for the observed outcome. Researchers and participants bring to the experiment a myriad of characteristics, some learned and others inherent. For example, sex, weight, hair, eye, and skin color, personality, mental capabilities, and physical abilities, but also attitudes like motivation or willingness to participate.

During the selection step of the research study, if an unequal number of test subjects have similar subject-related variables there is a threat to the internal validity. For example, a researcher created two test groups, the experimental and the control groups. The subjects in both groups are not alike with regard to the independent variable but similar in one or more of the subject-related variables.

Self-selection also has a negative effect on the interpretive power of the dependent variable. This occurs often in online surveys where individuals of specific demographics opt into the test at higher rates than other demographics.

History

Events outside of the study/experiment or between repeated measures of the dependent variable may affect participants' responses to experimental procedures. Often, these are large scale events (natural disaster, political change, etc.) that affect participants' attitudes and behaviors such that it becomes impossible to determine whether any change on the dependent measures is due to the independent variable, or the historical event.

Maturation

Subjects change during the course of the experiment or even between measurements. For example, young children might mature and their ability to concentrate may change as they grow up. Both permanent changes, such as physical growth and temporary ones like fatigue, provide "natural" alternative explanations; thus, they may change the way a subject would react to the independent variable. So upon completion of the study, the researcher may not be able to determine if the cause of the discrepancy is due to time or the independent variable.

Repeated Testing (Also Referred to as Testing Effects)

Repeatedly measuring the participants may lead to bias. Participants may remember the correct answers or may be conditioned to know that they are being tested. Repeatedly taking (the same or similar) intelligence tests usually leads to score gains, but instead of concluding that the underlying skills have changed for good, this threat to Internal Validity provides good rival hypotheses.

Instrument Change (Instrumentality)

The instrument used during the testing process can change the experiment. This also refers to observers being more concentrated or primed, or having unconsciously changed the criteria they use to make judgments. This can also be an issue with self-report measures given at different times. In this case the impact may be mitigated through the use of retrospective pretesting. If any instrumentation changes occur, the internal validity of the main conclusion is affected, as alternative explanations are readily available.

Regression Toward the Mean

This type of error occurs when subjects are selected on the basis of extreme scores (one far away from the mean) during a test. For example, when children with the worst reading scores are selected to participate in a reading course, improvements at the end of the course might be due to regression toward the mean and not the course's effectiveness. If the children had been tested again before the course started, they would likely have obtained better scores anyway. Likewise, extreme outliers on individual scores are more likely to be captured in one instance of testing but will likely evolve into a more normal distribution with repeated testing.

Mortality/Differential Attrition

This error occurs if inferences are made on the basis of only those participants that have participated from the start to the end. However, participants may have dropped out of the study before completion, and maybe even due to the study or programme or experiment itself. For example, the percentage of group members having quit smoking at post-test was found much higher in a group having received a quit-smoking training program than in the control group. However, in the experimental group only 60% have completed the program. If this attrition is systematically related to any feature of the study, the administration of the independent variable, the instrumentation, or if dropping out leads to relevant bias between groups, a whole class of alternative explanations is possible that account for the observed differences.

Selection-maturation Interaction

This occurs when the subject-related variables, color of hair, skin color, etc., and the time-related variables, age, physical size, etc., interact. If a discrepancy between the two groups occurs between the testing, the discrepancy may be due to the age differences in the age categories.

Diffusion

If treatment effects spread from treatment groups to control groups, a lack of differences between experimental and control groups may be observed. This does not mean, however, that the independent variable has no effect or that there is no relationship between dependent and independent variable.

Compensatory Rivalry/Resentful Demoralization

Behavior in the control groups may alter as a result of the study. For example, control group members may work extra hard to see that expected superiority of the experimental group is not demonstrated. Again, this does not mean that the independent variable produced no effect or that there is no relationship between dependent and independent variable. Vice versa, changes in the dependent variable may only be affected due to a demoralized control group, working less hard or motivated, not due to the independent variable.

Experimenter Bias

Experimenter bias occurs when the individuals who are conducting an experiment inadvertently affect the outcome by non-consciously behaving in different ways to members of control and experimental groups. It is possible to eliminate the possibility of experimenter bias through the use of double blind study designs, in which the experimenter is not aware of the condition to which a participant belongs.

For eight of these threats there exists the first letter mnemonic *THIS MESS*, which refers to the first letters of *T*esting (repeated testing), *H*istory, *I*nstrument change, *S*tatistical Regression toward the mean, *M*aturation, *E*xperimental mortality, *S*election and *S*election Interaction.

External Validity

External validity is the validity of generalized (causal) inferences in scientific research, usually based on experiments as experimental validity. In other words, it is the extent to which the results of a study can be generalized to other situations and to other people. Mathematical analysis of external validity concerns a determination of whether generalization across heterogeneous populations is feasible, and devising statistical and computational methods that produce valid generalizations.

Threats to External Validity

"A threat to external validity is an explanation of how you might be wrong in making a generalization." Generally, generalizability is limited when the cause (i.e. the independent variable) depends on other factors; therefore, all threats to external validity interact with the independent variable - a so-called background factor x treatment interaction.

- Aptitude–treatment Interaction: The sample may have certain features that may interact with the independent variable, limiting generalizability. For example, inferences based on comparative psychotherapy studies often employ specific samples (e.g. volunteers, highly depressed, no comorbidity). If psychotherapy is found effective for these sample patients, will it also be effective for non-volunteers or the mildly depressed or patients with concurrent other disorders?

- Situation: All situational specifics (e.g. treatment conditions, time, location, lighting, noise, treatment administration, investigator, timing, scope and extent of measurement, etc. etc.) of a study potentially limit generalizability.

- Pre-test effects: If cause-effect relationships can only be found when pre-tests are carried out, then this also limits the generality of the findings.

- Post-test effects: If cause-effect relationships can only be found when post-tests are carried out, then this also limits the generality of the findings.

- Reactivity (placebo, novelty, and Hawthorne effects): If cause-effect relationships are found they might not be generalizable to other settings or situations if the effects found only occurred as an effect of studying the situation.

- Rosenthal effects: Higher expectations may lead to better performance.

Cook and Campbell made the crucial distinction between generalizing *to* some population and generalizing *across* subpopulations defined by different levels of some background factor. Lynch has argued that it is almost never possible to generalize *to* meaningful populations except as a snapshot of history, but it is possible to test the degree to which the effect of some cause on some dependent variable generalizes *across* subpopulations that vary in some background factor. That requires a test of whether the treatment effect being investigated is moderated by interactions with one or more background factors.

Disarming Threats to External Validity

Whereas enumerating threats to validity may help researchers avoid unwarranted generalizations, many of those threats can be disarmed, or neutralized in a systematic way, so as to enable a valid generalization. Specifically, experimental findings from one population can be "re-processed", or "re-calibrated" so as to circumvent population differences and produce valid generalizations in a second population, where experiments cannot be performed. Pearl and Bareinboim classified generalization problems into two categories: (1) those that lend themselves to valid re-calibration, and (2) those where external validity is theoretically impossible. Using graph-based calculus, they derived a necessary and sufficient condition for a problem instance to enable a valid generalization, and devised algorithms that automatically produce the needed re-calibration, whenever such exists. This reduces the external validity problem to an exercise in graph theory, and has led some philosophers to conclude that the problem is now solved.

An important variant of the external validity problem deals with selection bias, also known as sampling bias— that is, bias created when studies are conducted on non-representative samples of the intended population. For example, if a clinical trial is conducted on college students, an investigator may wish to know whether the results generalize to the entire population, where attributes such as age, education, and income differ substantially from those of a typical student. The graph-based method of Bareinboim and Pearl identifies conditions under which sample selection bias can be circumvented and, when these conditions are met, the method constructs an unbiased estimator of the average causal effect in the entire population. The main difference between generalization from improperly sampled studies and generalization across disparate populations lies in the fact that disparities among populations are usually caused by preexisting factors, such as age or ethnicity, whereas selection bias is often caused by post-treatment conditions, for example, patients dropping out of the study, or patients selected by severity of injury. When selection is governed by post-treatment factors, unconventional re-calibration methods are required to ensure bias-free estimation, and these methods are readily obtained from the problem's graph.

Examples

If age is judged to be a major factor causing treatment effect to vary from individual to individual, then age differences between the sampled students and the general population would lead to a biased estimate of the average treatment effect in that population. Such bias can be corrected though by a simple re-weighing procedure: We take the age-specific effect in the student subpopulation and compute its average using the age distribution in the general population. This would give us an unbiased estimate of the average treatment effect in the population.

If, on the other hand, the relevant factor that distinguishes the study sample from the general population is in itself affected by the treatment, then a different re-weighing scheme need be invoked. Calling this factor *Z*, we again average the *z*-specific effect of *X* on *Y* in the experimental sample, but now we weigh it by the "causal effect" of *X* on *Z*. In other words, the new weight is the proportion of units attaining level *Z*=*z* had treatment *X*=*x* been administered to the entire population. This interventional probability, often written $P(Z = z \mid do(X = x))$, can sometimes be estimated from observational studies in the general population.

A typical example of this nature occurs when *Z* is a mediator between the treatment and outcome, For instance, the treatment may be a cholesterol- reducing drug, *Z* may be cholesterol level, and *Y* life expectancy. Here, *Z* is both affected by the treatment and a major factor in determining the outcome, *Y*. Suppose that subjects selected for the experimental study tend to have higher cholesterol levels than is typical in the general population. To estimate the average effect of the drug on survival in the entire population, we first compute the *z*-specific treatment effect in the experimental study, and then average it using $P(Z = z \mid do(X = x))$ as a weighting function. The estimate obtained will be bias-free even when *Z* and *Y* are confounded — that is, when there is an unmeasured common factor that affects both *Z* and *Y*.

The precise conditions ensuring the validity of this and other weighting schemes are formulated in Bareinboim and Pearl, 2016 and Bareinboim et al., 2014.

External, Internal, and Ecological Validity

In many studies and research designs, there may be a "trade-off" between internal validity and external validity: When measures are taken or procedures implemented aiming at increasing the chance for higher degrees of internal validity, these measures may also limit the generalizability of the findings. This situation has led many researchers call for "ecologically valid" experiments. By that they mean that experimental procedures should resemble "real-world" conditions. They criticize the lack of ecological validity in many laboratory-based studies with a focus on artificially controlled and constricted environments. Some researchers think external validity and ecological validity are closely related in the sense that causal inferences based on ecologically valid research designs often allow for higher degrees of generalizability than those obtained in an artificially produced lab environment. However, this again relates to the distinction between generalizing to some population (closely related to concerns about ecological validity) and generalizing across subpopulations that differ on some background factor. Some findings produced in ecologically valid research settings may hardly be generalizable, and some findings produced in highly controlled settings may claim near-universal external validity. Thus, external and ecological validity are independent – a study may possess external validity but not ecological validity, and vice versa.

Qualitative Research

Within the qualitative research paradigm, external validity is replaced by the concept of transferability. Transferability is the ability of research results to transfer to situations with similar parameters, populations and characteristics.

External Validity in Experiments

It is common for researchers to claim that experiments are by their nature low in external validity. Some claim that many drawbacks can occur when following the experimental method. By the virtue of gaining enough control over the situation so as to randomly assign people to conditions and rule out the effects of extraneous variables, the situation can become somewhat artificial and distant from real life.

There are two kinds of generalizability at issue:

1. The extent to which we can generalize from the situation constructed by an experimenter to real-life situations (*generalizability across situations*), and
2. The extent to which we can generalize from the people who participated in the experiment to people in general (*generalizability across people*)

However, both of these considerations pertain to Cook and Campbell's concept of generalizing *to* some target population rather than the arguably more central task of assessing the generalizability of findings from an experiment *across* subpopulations that differ from the specific situation studied and people who differ from the respondents studied in some meaningful way.

Critics of experiments suggest that external validity could be improved by use of field settings (or, at a minimum, realistic laboratory settings) and by use of true probability samples of respondents. However, if one's goal is to understand generalizability *across* subpopulations that differ in situational or personal background factors, these remedies do not have the efficacy in increasing external validity that is commonly ascribed to them. If background factor X treatment interactions exist of which the researcher is unaware (as seems likely), these research practices can mask a substantial lack of external validity. Dipboye and Flanagan (1979), writing about industrial and organizational psychology, note that the evidence is that findings from one field setting and from one lab setting are equally *unlikely* to generalize to a second field setting. Thus, field studies are not by their nature high in external validity and laboratory studies are not by their nature low in external validity. It depends in both cases whether the particular treatment effect studied would change with changes in background factors that are held constant in that study. If one's study is "unrealistic" on the level of some background factor that does not interact with the treatments, it has no effect on external validity. It is only if an experiment holds some background factor constant at an unrealistic level and if varying that background factor would have revealed a strong Treatment x Background factor interaction, that external validity is threatened.

Generalizability Across Situations

Research in psychology experiments attempted in universities are often criticized for being conducted in artificial situations and that it cannot be generalized to real life. To solve this problem, social psychologists attempt to increase the generalizability of their results by making their studies as realistic as possible. As noted above, this is in the hope of generalizing to some specific population. Realism per se does not help the make statements about whether the results would change if the setting were somehow more realistic, or if study participants were placed in a different realistic setting. If only one setting is tested, it is not possible to make statements about generalizability across settings.

However, many authors conflate external validity and realism. There is more than one way that an experiment can be realistic:

1. The similarity of an experimental situation to events that occur frequently in everyday life—it is clear that many experiments are decidedly unreal.
2. In many experiments, people are placed in situations they would rarely encounter in everyday life.

This is referred to the extent to which an experiment is similar to real-life situations as the experiment's mundane realism.

It is more important to ensure that a study is high in psychological realism—how similar the psychological processes triggered in an experiment are to psychological processes that occur in everyday life.

Psychological realism is heightened if people find themselves engrossed in a real event. To accomplish this, researchers sometimes tell the participants a cover story—a false description of the study's purpose. If however, the experimenters were to tell the participants the purpose of the experiment then such a procedure would be low in psychological realism. In everyday life, no one knows when emergencies are going to occur and people do not have time to plan responses to them. This means that the kinds of psychological processes triggered would differ widely from those of a real emergency, reducing the psychological realism of the study.

People don't always know why they do what they do, or what they do until it happens. Therefore, describing an experimental situation to participants and then asking them to respond normally will produce responses that may not match the behavior of people who are actually in the same situation. We cannot depend on people's predictions about what they would do in a hypothetical situation; we can only find out what people will really do when we construct a situation that triggers the same psychological processes as occur in the real world.

Generalizability Across People

Social psychologists study the way in which people in general are susceptible to social influence. Several experiments have documented an interesting, unexpected example of social influence, whereby the mere knowledge that others were present reduced the likelihood that people helped.

The only way to be certain that the results of an experiment represent the behaviour of a particular population is to ensure that participants are randomly selected from that population. Samples in experiments cannot be randomly selected just as they are in surveys because it is impractical and expensive to select random samples for social psychology experiments. It is difficult enough to convince a random sample of people to agree to answer a few questions over the telephone as part of a political poll, and such polls can cost thousands of dollars to conduct. Moreover, even if one somehow was able to recruit a truly random sample, there can be unobserved heterogeneity in the effects of the experimental treatments... A treatment can have a positive effect on some subgroups but a negative effect on others. The effects shown in the treatment averages may not generalize to any subgroup.

Many researchers address this problem by studying basic psychological processes that make people susceptible to social influence, assuming that these processes are so fundamental that they are universally shared. Some social psychologist processes do vary in different cultures and in those cases, diverse samples of people have to be studied.

Replications

The ultimate test of an experiment's external validity is replication — conducting the study over again, generally with different subject populations or in different settings. Researches will often use different methods, to see if they still get the same results.

When many studies of one problem are conducted, the results can vary. Several studies might find an effect of the number of bystanders on helping behaviour, whereas a few do not. To make sense out of this, there is a statistical technique called meta-analysis that averages the results of two or more studies to see if the effect of an independent variable is reliable. A meta analysis essentially tells us the probability that the findings across the results of many studies are attributable to chance or to the independent variable. If an independent variable is found to have an effect in only of 20 studies, the meta-analysis will tell you that that one study was an exception and that, on average, the independent variable is not influencing the dependent variable. If an independent variable is having an effect in most of the studies, the meta analysis is likely to tell us that, on average, it does influence the dependent variable.

There can be reliable phenomena that are not limited to the laboratory. For example, increasing the number of bystanders has been found to inhibit helping behaviour with many kinds of people, including children, university students, and future ministers; in Israel; in small towns and large cities in the U.S.; in a variety of settings, such as psychology laboratories, city streets, and subway trains; and with a variety of types of emergencies, such as seizures, potential fires, fights, and accidents, as well as with less serious events, such as having a flat tire. Many of these replications have been conducted in real-life settings where people could not possibly have known that an experiment was being conducted.

The Basic Dilemma of the Social Psychologist

When conducting experiments in psychology, some believe that there is always a trade-off between internal and external validity—

1. having enough control over the situation to ensure that no extraneous variables are influencing the results and to randomly assign people to conditions, and
2. ensuring that the results can be generalized to everyday life.

Some researchers believe that a good way to increase external validity is by conducting field experiments. In a field experiment, people's behavior is studied outside the laboratory, in its natural setting. A field experiment is identical in design to a laboratory experiment, except that it is conducted in a real-life setting. The participants in a field experiment are unaware that the events they experience are in fact an experiment. Some claim that the external validity of such an experiment is high because it is taking place in the real world, with real people who are more diverse than a typical university student sample. However, as real-world settings differ dramatically, findings in one real world setting may or may not generalize to another real world setting.

Neither internal nor external validity are captured in a single experiment. Social psychologists opt first for internal validity, conducting laboratory experiments in which people are randomly assigned to different conditions and all extraneous variables are controlled. Other social psychologists prefer external validity to control, conducting most of their research in field studies. And many do both. Taken together, both types of studies meet the requirements of the perfect experiment. Through replication, researchers can study a given research question with maximal internal and external validity.

References

- Mills, Albert J.; Gabrielle Durepos; Elden Wiebe. (Eds.). (2010). Encyclopedia of Case Study Research. Sage Publications. California. p. xxxi. ISBN 978-1-4129-5670-3.
- Robert K. Yin. Case Study Research: Design and Methods. 5th Edition. Sage Publications. California, 2014. Pages 5-6. ISBN 978-1-4522-4256-9
- Suzanne Corkin. Permanent Present Tense: The Unforgettable Life of the Amnesic Patient, H.M.. Basic Books. New York. 2013. ISBN 978-0-4650-3159-7
- Rodger Kessler & Dale Stafford. Editors. Collaborative Medicine Case Studies: Evidence in Practice. Springer. New York. 2008. ISBN 978-0-3877-6893-9
- Creswell, John (2009). Research Design; Qualitative and Quantitative and Mixed Methods Approaches. London: Sage. ISBN 978-1-4522-2609-5.
- Barney G. Glaser and Strauss, The discovery of grounded theory: Strategies for qualitative research (New York: Aldine, 1967). ISBN 978-0202302607
- W. Ellet. The Case Study Handbook: How to Read, Write, and Discuss Persuasively about Cases. Harvard Business School Press. Boston, MA. 2007. ISBN 978-1-422-10158-2
- Palmer, Grier; Iordanou, Ioanna (2015). Exploring Cases Using Emotion, Open Space and Creativity. Case-based Teaching and Learning for the 21st Century. Libri. pp. 19–38. ISBN 978 1 909818 57 6.
- Dinardo, J. (2008). "natural experiments and quasi-natural experiments". The New Palgrave Dictionary of Economics. pp. 856–859. doi:10.1057/9780230226203.1162. ISBN 978-0-333-78676-5.
- Rossi, Peter Henry; Mark W. Lipsey; Howard E. Freeman (2004). Evaluation: A Systematic Approach (7th ed.). SAGE. p. 237. ISBN 978-0-7619-0894-4.
- Shadish; Cook; Cambell (2002). Experimental and Quasi-Experimental Designs for Generalized Causal Inference. Boston: Houghton Mifflin. ISBN 0-395-61556-9.
- Campbell, D. T. (1988). Methodology and epistemology for social science: selected papers. University of Chicago Press. ISBN 0-226-09248-8.
- Croxton, Frederick Emory; Cowden, Dudley Johnstone; Klein, Sidney (1968) Applied General Statistics, Pitman. ISBN 9780273403159 (page 625)
- Dietrich, Cornelius Frank (1991) Uncertainty, Calibration and Probability: The Statistics of Scientific and Industrial Measurement 2nd Edition, A. Higler. ISBN 9780750300605.
- Dales, Richard C. (1973), The Scientific Achievement of the Middle Ages (The Middle Ages Series), University of Pennsylvania Press, ISBN 9780812210576
- di Francia, G. Toraldo (1981), The Investigation of the Physical World, Cambridge University Press, ISBN 0-521-29925-X.

3

Social and Academic Research: An Integrated Study

Social research is a research that is conducted by students or social scientists. The topics explained in the section are quantitative research, multimethodology, academic publishing, peer review, clinical peer review etc. The aspects elucidated in this chapter are of vital importance, and provide a better understanding of social and academic research.

Social Research

Social research is research conducted by social scientists following a systematic plan. Social research methodologies can be classified along a quantitative/qualitative dimension.

- Quantitative designs approach social phenomena through quantifiable evidence, and often rely on statistical analysis of many cases (or across intentionally designed treatments in an experiment) to create valid and reliable general claims. Related to quantity.
- Qualitative designs emphasize understanding of social phenomena through direct observation, communication with participants, or analysis of texts, and may stress contextual subjective accuracy over generality. Related to quality.

While methods may be classified as quantitative or qualitative, most methods contain elements of both. For example, qualitative data analysis often involves a fairly structured approach to coding the raw data into systematic information, and quantifying intercoder reliability. Thus, there is often a more complex relationship between "qualitative" and "quantitative" approaches than would be suggested by drawing a simple distinction between them.

Social scientists employ a range of methods in order to analyse a vast breadth of social phenomena: from census survey data derived from millions of individuals, to the in-depth analysis of a single agent's social experiences; from monitoring what is happening on contemporary streets, to the investigation of ancient historical documents. Methods rooted in classical sociology and statistics have formed the basis for research in other disciplines, such as political science, media studies, program evaluation and market research.

Methodology

Social scientists are divided into camps of support for particular research techniques. These disputes relate to the historical core of social theory (positivism and antipositivism; structure and agency). While very different in many aspects, both qualitative and quantitative approaches involve a systematic interaction between theory and data. The choice of method often depends largely on what the researcher intends to investigate. For example, a researcher concerned with drawing a statistical generalization across an entire population may administer a survey questionnaire to a representative sample population. By contrast, a researcher who seeks full contextual understanding of an individuals' social actions may choose ethnographic participant observation or open-ended interviews. Studies will commonly combine, or 'triangulate', quantitative *and* qualitative methods as part of a 'multi-strategy' design.

Sampling

Typically a population is very large, making a census or a complete enumeration of all the values in that population infeasible. A 'sample' thus forms a manageable subset of a population. In positivist research, statistics derived from a sample are analysed in order to draw inferences regarding the population as a whole. The process of collecting information from a sample is referred to as 'sampling'. Sampling methods may be either 'random' (random sampling, systematic sampling, stratified sampling, cluster sampling) or non-random/nonprobability (convenience sampling, purposive sampling, snowball sampling). The most common reason for sampling is to obtain information about a population. Sampling is quicker and cheaper than a complete census of a population.

Methodological Assumptions

Social research is based on logic and empirical observations. Charles C. Ragin writes in his *Constructing Social Research* book that "Social research involved the interaction between ideas and evidence. Ideas help social researchers make sense of evidence, and researchers use evidence to extend, revise and test ideas." Social research thus attempts to create or validate theories through data collection and data analysis, and its goal is exploration, description, explanation, and prediction. It should never lead or be mistaken with philosophy or belief. Social research aims to find social patterns of regularity in social life and usually deals with social groups (aggregates of individuals), not individuals themselves (although science of psychology is an exception here). Research can also be divided into pure research and applied research. Pure research has no application on real life, whereas applied research attempts to influence the real world.

There are no laws in social science that parallel the laws in natural science. A law in social science is a universal generalization about a class of facts. A fact is an observed phenomenon, and observation means it has been seen, heard or otherwise experienced by researcher. A theory is a systematic explanation for the observations that relate to a particular aspect of social life. Concepts are the basic building blocks of theory and are abstract elements representing classes of phenomena. Axioms or postulates are basic assertions assumed to be true. Propositions are conclusions drawn about the relationships among concepts, based on analysis of axioms. Hypotheses are specified expectations about empirical reality derived from propositions. Social research involves testing these hypotheses to see if they are true.

Social research involves creating a theory, operationalization (measurement of variables) and observation (actual collection of data to test hypothesized relationship). Social theories are written in the language of variables, in other words, theories describe logical relationships between variables. Variables are logical sets of attributes, with people being the "carriers" of those variables (for example, gender can be a variable with two attributes: male and female). Variables are also divided into independent variables (data) that influences the dependent variables (which scientists are trying to explain). For example, in a study of how different dosages of a drug are related to the severity of symptoms of a disease, a measure of the severity of the symptoms of the disease is a dependent variable and the administration of the drug in specified doses is the independent variable. Researchers will compare the different values of the dependent variable (severity of the symptoms) and attempt to draw conclusions.

Guidelines for "Good Research"

When social scientists speak of "good research" the guidelines refer to how the science is mentioned and understood. It does not refer to how what the results are but how they are figured. Glenn Firebaugh summarizes the principles for good research in his book *Seven Rules for Social Research*. The first rule is that "There should be the possibility of surprise in social research." As Firebaugh (p. 1) elaborates: "Rule 1 is intended to warn that you don't want to be blinded by preconceived ideas so that you fail to look for contrary evidence, or you fail to recognize contrary evidence when you do encounter it, or you recognize contrary evidence but suppress it and refuse to accept your findings for what they appear to say."

In addition, good research will "look for differences that make a difference" (Rule 2) and "build in reality checks" (Rule 3). Rule 4 advises researchers to replicate, that is, "to see if identical analyses yield similar results for different samples of people" (p. 90). The next two rules urge researchers to "compare like with like" (Rule 5) and to "study change" (Rule 6); these two rules are especially important when researchers want to estimate the effect of one variable on another (e.g. how much does college education actually matter for wages?). The final rule, "Let method be the servant, not the master," reminds researchers that methods are the means, not the end, of social research; it is critical from the outset to fit the research design to the research issue, rather than the other way around.

Explanations in social theories can be idiographic or nomothetic. An idiographic approach to an explanation is one where the scientists seek to exhaust the idiosyncratic causes of a particular condition or event, i.e. by trying to provide all possible explanations of a particular case. Nomothetic explanations tend to be more general with scientists trying to identify a few causal factors that impact a wide class of conditions or events. For example, when dealing with the problem of how people choose a job, idiographic explanation would be to list all possible reasons why a given person (or group) chooses a given job, while nomothetic explanation would try to find factors that determine why job applicants in general choose a given job.

Research in science and in social science is a long, slow and difficult process that sometimes produces false results because of methodological weaknesses and in rare cases because of fraud, so that reliance on any one study is inadvisable.

Ethics

The ethics of social research are shared with those of medical research. In the United States, these are formalized by the Belmont report as:

Respect for Persons

The principle of respect for persons holds that (a) individuals should be respected as autonomous agents capable of making their own decisions, and that (b) subjects with diminished autonomy deserve special considerations. A cornerstone of this principle is the use of informed consent.

Beneficence

The principle of beneficence holds that (a) the subjects of research should be protected from harm, and, (b) the research should bring tangible benefits to society. By this definition, research with no scientific merit is automatically considered unethical.

Justice

The principle of justice states the benefits of research should be distributed fairly. The definition of fairness used is case-dependent, varying between "(1) to each person an equal share, (2) to each person according to individual need, (3) to each person according to individual effort, (4) to each person according to societal contribution, and (5) to each person according to merit."

Types of Method

The following list of research methods is not exhaustive:

Statistical–quantitative methods	Qualitative methods	Mixed methods
• Cluster analysis • Correlation and association • Multivariate statistics • Regression analysis • Social network analysis • Social sequence analysis • Surveys and questionnaire • Structural equation modeling • Survey research • Quantitative marketing research	• Analytic induction • Case study • Ethnography • Life history • Morphological analysis • Most significant change technique • Participant observation • Textual analysis • Unstructured interview	• Archival research • Content analysis • Longitudinal study • Focus group • Historical method • Semi-structured interview • Structured interview • Triangulation (social science)

Foundations of Social Research

Sociological Positivism

The origin of the survey can be traced back at least early as the Domesday Book in 1086, while some scholars pinpoint the origin of demography to 1663 with the publication of John Graunt's *Natural and Political Observations upon the Bills of Mortality*. Social research began most intentionally, however, with the positivist philosophy of science in the early 19th century.

Statistical sociological research, and indeed the formal academic discipline of sociology, began with the work of Émile Durkheim (1858–1917). While Durkheim rejected much of the detail of Comte's philosophy, he retained and refined its method, maintaining that the social sciences are a logical continuation of the natural ones into the realm of human activity, and insisting that they may retain the same objectivity, rationalism, and approach to causality. Durkheim set up the first European department of sociology at the University of Bordeaux in 1895, publishing his Rules of the Sociological Method (1895). In this text he argued: "[o]ur main goal is to extend scientific rationalism to human conduct.... What has been called our positivism is but a consequence of this rationalism."

Émile Durkheim

Durkheim's seminal monograph, *Suicide* (1897), a case study of suicide rates among Catholic and Protestant populations, distinguished sociological analysis from psychology or philosophy. By carefully examining suicide statistics in different police districts, he attempted to demonstrate that Catholic communities have a lower suicide rate than that of Protestants, something he attributed to social (as opposed to individual or psychological) causes. He developed the notion of objective *suis generis* "social facts" to delineate a unique empirical object for the science of sociology to study. Through such studies he posited that sociology would be able to determine whether any given society is 'healthy' or 'pathological', and seek social reform to negate organic breakdown or "social anomie". For Durkheim, sociology could be described as the "science of institutions, their genesis and their functioning".

Modern Methodologies

In the mid-20th century there was a general—but not universal—trend for U.S.American sociology to be more scientific in nature, due to the prominence at that time of action theory and other system-theoretical approaches. Robert K. Merton released his *Social Theory and Social Structure* (1949). By the turn of the 1960s, sociological research was increasingly employed as a tool by governments and businesses worldwide. Sociologists developed new types of quantitative and qualitative research methods. Paul Lazarsfeld founded Columbia University's Bureau of Applied Social Research, where he exerted a tremendous influence over the techniques and the organization of social research. His many contributions to sociological method have earned him the title of the "founder of modern empirical sociology". Lazarsfeld made great strides in statistical survey analysis, panel methods, latent structure analysis, and contextual analysis. Many of his ideas have been so influential as to now be considered self-evident.

Quantitative Research

In natural sciences and social sciences, quantitative research is the systematic empirical investigation of observable phenomena via statistical, mathematical or computational techniques. The objective of quantitative research is to develop and employ mathematical models, theories and/or hypotheses pertaining to phenomena. The process of measurement is central to quantitative research because it provides the fundamental connection between empirical observation and mathematical expression of quantitative relationships. Quantitative data is any data that is in numerical form such as statistics, percentages, etc. The researcher analyzes the data with the help of statistics. The researcher is hoping the numbers will yield an unbiased result that can be generalized to some larger population. Qualitative research, on the other hand, asks broad questions and collects word data from phenomena or participants. The researcher looks for themes and describes the information in themes and patterns exclusive to that set of participants.

In social sciences, quantitative research is widely used in psychology, economics, demography, sociology, marketing, community health, health & human development, gender and political science, and less frequently in anthropology and history. Research in mathematical sciences such as physics is also 'quantitative' by definition, though this use of the term differs in context. In the social sciences, the term relates to empirical methods, originating in both philosophical positivism and the history of statistics, which contrast with qualitative research methods.

Qualitative research produces information only on the particular cases studied, and any more general conclusions are only hypotheses. Quantitative methods can be used to verify which of such hypotheses are true.

A comprehensive analysis of 1274 articles published in the top two American sociology journals between 1935 and 2005 found that roughly two thirds of these articles used quantitative methods.

Overview

Quantitative research is generally made using scientific methods, which can include:

- The generation of models, theories and hypotheses

- The development of instruments and methods for measurement
- Experimental control and manipulation of variables
- Collection of empirical data
- Modeling and analysis of data

Quantitative research is often contrasted with qualitative research, which is the examination, analysis and interpretation of observations for the purpose of discovering underlying meanings and patterns of relationships, including classifications of types of phenomena and entities, in a manner that does not involve mathematical models. Approaches to quantitative psychology were first modeled on quantitative approaches in the physical sciences by Gustav Fechner in his work on psychophysics, which built on the work of Ernst Heinrich Weber. Although a distinction is commonly drawn between qualitative and quantitative aspects of scientific investigation, it has been argued that the two go hand in hand. For example, based on analysis of the history of science, Kuhn concludes that "large amounts of qualitative work have usually been prerequisite to fruitful quantification in the physical sciences". Qualitative research is often used to gain a general sense of phenomena and to form theories that can be tested using further quantitative research. For instance, in the social sciences qualitative research methods are often used to gain better understanding of such things as intentionality (from the speech response of the researchee) and meaning (why did this person/group say something and what did it mean to them?) (Kieron Yeoman).

Although quantitative investigation of the world has existed since people first began to record events or objects that had been counted, the modern idea of quantitative processes have their roots in Auguste Comte's positivist framework. Positivism emphasized the use of the scientific method through observation to empirically test hypotheses explaining and predicting what, where, why, how, and when phenomena occurred. Positivist scholars like Comte believed only scientific methods rather than previous spiritual explanations for human behavior could advance.

Use of Statistics

Statistics is the most widely used branch of mathematics in quantitative research outside of the physical sciences, and also finds applications within the physical sciences, such as in statistical mechanics. Statistical methods are used extensively within fields such as economics, social sciences and biology. Quantitative research using statistical methods starts with the collection of data, based on the hypothesis or theory. Usually a big sample of data is collected – this would require verification, validation and recording before the analysis can take place. Software packages such as SPSS and R are typically used for this purpose. Causal relationships are studied by manipulating factors thought to influence the phenomena of interest while controlling other variables relevant to the experimental outcomes. In the field of health, for example, researchers might measure and study the relationship between dietary intake and measurable physiological effects such as weight loss, controlling for other key variables such as exercise. Quantitatively based opinion surveys are widely used in the media, with statistics such as the proportion of respondents in favor of a position commonly reported. In opinion surveys, respondents are asked a set of structured questions and their responses are tabulated. In the field of climate science, researchers compile and compare statistics such as temperature or atmospheric concentrations of carbon dioxide.

Empirical relationships and associations are also frequently studied by using some form of general linear model, non-linear model, or by using factor analysis. A fundamental principle in quantitative research is that correlation does not imply causation, although some such as Clive Granger suggest that a series of correlations can imply a degree of causality. This principle follows from the fact that it is always possible a spurious relationship exists for variables between which covariance is found in some degree. Associations may be examined between any combination of continuous and categorical variables using methods of statistics.

Measurement

Views regarding the role of measurement in quantitative research are somewhat divergent. Measurement is often regarded as being only a means by which observations are expressed numerically in order to investigate causal relations or associations. However, it has been argued that measurement often plays a more important role in quantitative research. For example, Kuhn argued that within quantitative research, the results that are shown can prove to be strange. This is because accepting a theory based on results of quantitative data could prove to be a natural phenomenon. He argued that such abnormalities are interesting when done during the process of obtaining data, as seen below:

> When measurement departs from theory, it is likely to yield mere numbers, and their very neutrality makes them particularly sterile as a source of remedial suggestions. But numbers register the departure from theory with an authority and finesse that no qualitative technique can duplicate, and that departure is often enough to start a search (Kuhn, 1961, p. 180).

In classical physics, the theory and definitions which underpin measurement are generally deterministic in nature. In contrast, probabilistic measurement models known as the Rasch model and Item response theory models are generally employed in the social sciences. Psychometrics is the field of study concerned with the theory and technique for measuring social and psychological attributes and phenomena. This field is central to much quantitative research that is undertaken within the social sciences.

Quantitative research may involve the use of *proxies* as stand-ins for other quantities that cannot be directly measured. Tree-ring width, for example, is considered a reliable proxy of ambient environmental conditions such as the warmth of growing seasons or amount of rainfall. Although scientists cannot directly measure the temperature of past years, tree-ring width and other climate proxies have been used to provide a semi-quantitative record of average temperature in the Northern Hemisphere back to 1000 A.D. When used in this way, the proxy record (tree ring width, say) only reconstructs a certain amount of the variance of the original record. The proxy may be calibrated (for example, during the period of the instrumental record) to determine how much variation is captured, including whether both short and long term variation is revealed. In the case of tree-ring width, different species in different places may show more or less sensitivity to, say, rainfall or temperature: when reconstructing a temperature record there is considerable skill in selecting proxies that are well correlated with the desired variable.

Relationship with Qualitative Methods

In most physical and biological sciences, the use of either quantitative or qualitative methods is uncontroversial, and each is used when appropriate. In the social sciences, particularly in sociology, social anthropology and psychology, the use of one or other type of method can be a matter of controversy and even ideology, with particular schools of thought within each discipline favouring one type of method and pouring scorn on to the other. The majority tendency throughout the history of social science, however, is to use eclectic approaches-by combining both methods. Qualitative methods might be used to understand the meaning of the conclusions produced by quantitative methods. Using quantitative methods, it is possible to give precise and testable expression to qualitative ideas. This combination of quantitative and qualitative data gathering is often referred to as mixed-methods research.

Examples

- Research that consists of the percentage amounts of all the elements that make up Earth's atmosphere.
- Survey that concludes that the average patient has to wait two hours in the waiting room of a certain doctor before being selected.
- An experiment in which group x was given two tablets of Aspirin a day and Group y was given two tablets of a placebo a day where each participant is randomly assigned to one or other of the groups. The numerical factors such as two tablets, percent of elements and the time of waiting make the situations and results quantitative.
- In finance, quantitative research into the stock markets is used to develop models to price complex trades, and develop algorithms to exploit investment hypotheses, as seen in quantitative hedge funds and Trading Strategy Indices.

Multimethodology

Multimethodology or multimethod research includes the use of more than one method of data collection or research in a research study or set of related studies. Mixed methods research is more specific in that it includes the mixing of qualitative and quantitative data, methods, methodologies, and/or paradigms in a research study or set of related studies. One could argue that mixed methods research is a special case of multimethod research. Another applicable, but less often used label, for multi or mixed research is methodological pluralism. All of these approaches to professional and academic research emphasize that monomethod research can be improved through the use of multiple data, methods, methodologies, perspectives, standpoints, and paradigms.

The term 'multimethodology' was used starting in the 1980s and in the 1989 book Multimethod Research: A Synthesis of Styles by John Brewer and Albert Hunter. During the 1990s and currently, the term 'mixed methods research' has become more popular for this research movement in the behavioral, social, business, and health sciences. This pluralistic research approach has been gaining in popularity since the 1980s.

Multi and Mixed Methods Research

There are three broad classes of research studies that are currently being labeled "mixed methods research" (Johnson, Onwuegbuzie, & Turner, 2007)

1. *Quantitatively driven approaches/designs* in which the research study is, at its core, a quantitative study with qualitative data/method added to supplement and improve the quantitative study by providing an added value and deeper, wider, and fuller or more complex answers to research questions; quantitative quality criteria are emphasized but high quality qualitative data also must be collected and analyzed;

2. *Qualitatively driven approaches/designs* in which the research study is, at its core, a qualitative study with quantitative data/method added to supplement and improve the qualitative study by providing an added value and deeper, wider, and fuller or more complex answers to research questions; qualitative quality criteria are emphasized but high quality quantitative data also must be collected and analyzed;

3. *Interactive or equal status designs* in which the research study equally emphasizes (interactively and through integration) quantitative and qualitative data, methods, methodologies, and paradigms. This third design is often done through the use of a team composed of an expert in quantitative research, an expert in qualitative research, and an expert in mixed methods research to help with dialogue and continual integration. In this type of mixed study, quantita-

> tive and qualitative and mixed methods quality criteria are emphasized. This use of multiple quality criteria is seen in the concept of *multiple validities legitimation* (Onwuegbuzie & Johnson, 2006; Johnson & Christensen, 2014). Here is a definition of this important type of validity or legitimation: Multiple validities legitimation "refers to the extent to which the mixed methods researcher successfully addresses and resolves all relevant validity types, including the quantitative and qualitative validity types discussed earlier in this chapter as well as the mixed validity dimensions. In other words, the researcher must identify and address all of the relevant validity issues facing a particular research study. Successfully addressing the pertinent validity issues will help researchers produce the kinds of inferences and meta-inferences that should be made in mixed research"(Johnson & Christensen, 2014; page 311).

One major similarity between mixed methodologies and qualitative and quantitative taken separately is that researchers need to maintain focus on the original purpose behind their methodological choices. A major difference between the two however, is the way some authors differentiate the two, proposing that there is logic inherent in one that is different from the other. Creswell (2009) points out that in a quantitative study the researcher starts with a problem statement, moving on to the hypothesis and null hypothesis, through the instrumentation into a discussion of data collection, population, and data analysis. Creswell proposes that for a qualitative study the flow of logic begins with the purpose for the study, moves through the research questions discussed as data collected from a smaller group and then voices how they will be analysed.

A research *strategy* is a procedure for achieving a particular intermediary research objective — such as sampling, data collection, or data analysis. We may therefore speak of sampling strategies or data analysis strategies. The use of multiple strategies to enhance construct validity (a form of methodological triangulation) is now routinely advocated by methodologists. In short, mixing or integrating research strategies (qualitative and/or quantitative) in any and all research undertaking is now considered a common feature of good research.

A research *approach* refers to an integrated set of research principles and general procedural guidelines. Approaches are broad, holistic (but general) methodological guides or roadmaps that are associated with particular research motives or analytic interests. Two examples of analytic interests are population frequency distributions and prediction. Examples of research approaches include experiments, surveys, correlational studies, ethnographic research, and phenomenological inquiry. Each approach is ideally suited to addressing a particular analytic interest. For instance, experiments are ideally suited to addressing nomothetic explanations or probable cause; surveys — population frequency descriptions, correlations studies — predictions; ethnography — descriptions and interpretations of cultural processes; and phenomenology — descriptions of the essence of phenomena or lived experiences.

In a *single approach design* (*SAD*)(also called a "monomethod design") only one analytic interest is pursued. In a *mixed or multiple approach design (MAD)* two or more analytic interests are pursued. NOTE: a multiple approach design may include entirely "quantitative" approaches such as combining a survey and an experiment; or entirely "qualitative" approaches such as combining an ethnographic and a phenomenological inquiry, and a mixed approach design includes a mixture of the above (e.g., a mixture of quantitative and qualitative data, methods, methodologies, and/or paradigms).

A word of caution about the term "multimethodology". It has become quite common place to use the terms "method" and "methodology" as synonyms (as is the case with the above entry). However, there are convincing philosophical reasons for distinguishing the two. "Method" connotes a way of doing something — a procedure (such as a method of data collection). "Methodology" connotes a discourse about methods — i.e., a discourse about the adequacy and appropriateness of particular combination of research principles and procedures. The terms methodology and biology share a common suffix "*logy*." Just as bio-*logy* is a discourse about life — all kinds of life; so too, methodo-*logy* is a discourse about methods — all kinds of methods. It seems unproductive, therefore, to speak of multi-biologies or of multi-methodologies. It is very productive, however, to speak of multiple biological perspectives or of multiple methodological perspectives.

Desirability

The case for multimethodology or mixed methods research as a strategy for intervention and/or research is based on four observations:

1. Narrow views of the world are often misleading, so approaching a subject from different perspectives or paradigms may help to gain a holistic perspective.
2. There are different levels of social research (i.e.: biological, cognitive, social, etc.), and different methodologies may have particular strengths with respect to one of these levels. Using more than one should help to get a clearer picture of the social world and make for more adequate explanations.
3. Many existing practices already combine methodologies to solve particular problems, yet they have not been theorised sufficiently.
4. Multimethodology fits well with pragmatism.

Feasibility

There are also some hazards to multimethodological or mixed methods research approaches. Some of these problems include:

1. Many paradigms are at odds with each other. However, once the understanding of the difference is present, it can be an advantage to see many sides, and possible solutions may present themselves.

2. Cultural issues affect world views and analyzability. Knowledge of a new paradigm is not enough to overcome potential biases; it must be learned through practice and experience.

3. People have cognitive abilities that predispose them to particular paradigms. The logical thinker can more easily understand and use quantitative methodologies. It is easier to move from quantitative to qualitative, and not the reverse.

Software

A few qualitative research analysis software applications support some degree of quantitative integration, and the following software or web applications focus on mixed methods research:

- Dedoose is a web-based qualitative analysis application and mixed methods research tool developed by professors from UCLA, and is the successor to EthnoNotes.
- MAXQDA is a qualitative data analysis and mixed methods software developed by VERBI Software. Consult. Sozialforschung GmbH. Mixed methods functionality includes guided creation for Joint Displays.
- NVivo is qualitative and mixed methods data analysis software developed by QSR International.
- QDA Miner is a qualitative data analysis and mixed methods software developed by Provalis Research.
- Quirkos
- Raven's Eye is an online hybrid and multimethod natural language analysis software program that facilitates both qualitative and quantitative data analysis

Conclusion

Multimethodology and mixed methods research are desirable and feasible because they provide a more complete view, and because the requirement during the different phases of an intervention (or research project) make very specific demands on a general methodology. While it is demanding, it is more effective to choose the right tool for the job at hand.

It can be used when you want to build from one phase of research to another. You may first want to explore the data qualitatively to identify help in the development an instrument or to identify concepts/variables to test in a later quantitative study or phase of a single study. You engage in a mixed methods study when you want to construct a quantitatively-driven design, a qualitatively-driven design, or an interactive/equal-status design. Each of these come with advantages and disadvantages. For more information on designing multiple and mixed methods research studies see the following design typologies and other (anti-typology): Brewer and Hunter, 2006); Creswell & Plano Clark (2011); Greene (2007); Guest (2013); Johnson and Christensen (2014); Morgan, (2014); Morse and Niehaus (2009), and Teddlie and Tashakkori (2009).

Criticisms

Multimethodology is criticized by the adherents of incompatibility thesis - particularly post-structuralist and post-modernists. Its critics argue that mixed methods research is inherently wrong because quantitative and qualitative research paradigms should not be mixed.

Academic Publishing

Academic publishing is the subfield of publishing which distributes academic research and scholarship. Most academic work is published in academic journal article, book or thesis form. The part of academic written output that is not formally published but merely printed up or posted on the Internet is often called "grey literature". Most scientific and scholarly journals, and many academic and scholarly books, though not all, are based on some form of peer review or editorial refereeing to qualify texts for publication. Peer review quality and selectivity standards vary greatly from journal to journal, publisher to publisher, and field to field.

Most established academic disciplines have their own journals and other outlets for publication, although many academic journals are somewhat interdisciplinary, and publish work from several distinct fields or subfields. There is also a tendency for existing journals to divide into specialized sections as the field itself becomes more specialized. Along with the variation in review and publication procedures, the kinds of publications that are accepted as contributions to knowledge or research differ greatly among fields and subfields.

Academic publishing is undergoing major changes, as it makes the transition from the print to the electronic format. Business models are different in the electronic environment. Since the early 1990s, licensing of electronic resources, particularly journals, has been very common. Currently, an important trend, particularly with respect to journals in the sciences, is open access via the Internet. In open access publishing a journal article is made available free for all on the web by the publisher at the time of publication. It is typically made possible after the author pays hundreds or thousands of dollars in publication fees, thereby shifting the costs from the reader to the researcher or their funder. The Internet has facilitated open access self-archiving, in which authors themselves make a copy of their published articles available free for all on the web.

History

One of the earliest research journals is the *Philosophical Transactions of the Royal Society*, created in the 17th century. At that time, the act of publishing academic inquiry was controversial, and widely ridiculed. It was not at all unusual for a new discovery to be announced as an anagram, reserving priority for the discoverer, but indecipherable for anyone not in on the secret: both Isaac Newton and Leibniz used this approach. However, this method did not work well. Robert K. Merton, a sociologist, found that 92% of cases of simultaneous discovery in the 17th century ended in dispute. The number of disputes dropped to 72% in the 18th century, 59% by the latter half of the 19th century, and 33% by the first half of the 20th century. The decline in contested claims for priority in research discoveries can be credited to the increasing acceptance of the publication of papers in modern academic journals, with estimates suggesting that around 50 million journal articles have been published since the first appearance of the *Philosophical Transactions*.

The Royal Society was steadfast in its not yet popular belief that science could only move forward through a transparent and open exchange of ideas backed by experimental evidence.

Its content included obituaries of famous men, church history, and legal reports. The first issue appeared as a twelve-page quarto pamphlet on Monday, 5 January 1665. This was shortly before the first appearance of the *Philosophical Transactions of the Royal Society*, on 6 March 1665.

Publishers and Business Aspects

In the 1960s and 1970s, commercial publishers began to selectively acquire "top-quality" journals which were previously published by nonprofit academic societies. Due to the inelastic demand for these journals, the commercial publishers lost little of the market when they raised the prices significantly. Although there are over 2,000 publishers, as of 2013, five for-profit companies (Reed Elsevier, Springer Science+Business Media, Wiley-Blackwell, Taylor & Francis, and Sage) accounted for 50% of articles published. (Since 2013, Springer Science+Business Media has undergone a merger to form an even bigger company named Springer Nature.) Available data indicate that these companies have high profit margins, especially compared to the smaller publishers which likely operate with low margins. These factors have contributed to the "serials crisis" – from 1986 to 2005, the number of serials purchased has increased an average of 1.9% per year while total expenditures on serials has increased 7.6% per year.

Unlike most industries, in academic publishing the two most important inputs are provided "virtually free of charge". These are the articles and the peer review process. Publishers argue that they add value to the publishing process through support to the peer review group, including stipends, as well as through typesetting, printing, and web publishing. Investment analysts, however, have been skeptical of the value added by for-profit publishers, as exemplified by a 2005 Deutsche Bank analysis which stated that "we believe the publisher adds relatively little value to the publishing process... We are simply observing that if the process really were as complex, costly and value-added as the publishers protest that it is, 40% margins wouldn't be available."

Crisis

A crisis in academic publishing is "widely perceived"; the apparent crisis has to do with the combined pressure of budget cuts at universities and increased costs for journals (the serials crisis). The university budget cuts have reduced library budgets and reduced subsidies to university-affiliated publishers. The humanities have been particularly affected by the pressure on university publishers, which are less able to publish monographs when libraries can't afford to purchase them. For example, the ARL found that in "1986, libraries spent 44% of their budgets on books compared with 56% on journals; twelve years later, the ratio had skewed to 28% and 72%." Meanwhile, monographs are increasingly expected for tenure in the humanities. The Modern Language Association has expressed hope that electronic publishing will solve the issue.

In 2009 and 2010, surveys and reports found that libraries faced continuing budget cuts, with one survey in 2009 finding that one-third of libraries had their budgets cut by 5% or more.

Academic Journal Publishing Reform

Several models are being investigated such as open publication models or adding community-oriented features. It is also considered that "Online scientific interaction outside the traditional journal space is becoming more and more important to academic communication". In addition, experts have suggested measures to make the publication process more efficient in disseminating new and important findings by evaluating the worthiness of publication on the basis of the significance and novelty of the research finding.

Scholarly Paper

In academic publishing, a paper is an academic work that is usually published in an academic journal. It contains original research results or reviews existing results. Such a paper, also called an article, will only be considered valid if it undergoes a process of peer review by one or more *referees* (who are academics in the same field) who check that the content of the paper is suitable for publication in the journal. A paper may undergo a series of reviews, revisions and re-submissions before finally being accepted or rejected for publication. This process typically takes several months. Next there is often a delay of many months (or in some subjects, over a year) before an accepted manuscript appears. This is particularly true for the most popular journals where the number of accepted articles often outnumbers the space for printing. Due to this, many academics self-archive a 'pre-print' copy of their paper for free download from their personal or institutional website.

Some journals, particularly newer ones, are now published in electronic form only. Paper journals are now generally made available in electronic form as well, both to individual subscribers, and to libraries. Almost always these electronic versions are available to subscribers immediately upon publication of the paper version, or even before; sometimes they are also made available to non-subscribers, either immediately or after an embargo of anywhere from two to twenty-four months or more, in order to protect against loss of subscriptions. Journals having this delayed availability are sometimes called delayed open access journals. Ellison has reported that in economics the dramatic increase in opportunities to publish results online has led to a decline in the use of peer reviewed articles.

Categories of Papers

An academic paper typically belongs to some particular category such as:

- Research paper
- Case report or Case series
- Position paper
- Review article or Survey paper
- Species paper
- Technical paper

Note: Law review is the generic term for a journal of legal scholarship in the United States, often operating by rules radically different from those for most other academic journals.

Peer Review

Peer review is a central concept for most academic publishing; other scholars in a field must find a work sufficiently high in quality for it to merit publication. A secondary benefit of the process is an indirect guard against plagiarism, since reviewers are usually familiar with the sources consulted by the author(s). The origins of routine peer review for submissions dates to 1752 when the Royal Society of London took over official responsibility for *Philosophical Transactions*. However, there were some earlier examples.

While journal editors largely agree the system is essential to quality control in terms of rejecting poor quality work, there have been examples of important results that are turned down by one journal before being taken to others. Rena Steinzor wrote:

Perhaps the most widely recognized failing of peer review is its inability to ensure the identification of high-quality work. The list of important scientific papers that were initially rejected by peer-reviewed journals goes back at least as far as the editor of *Philosophical Transaction's* 1796 rejection of Edward Jenner's report of the first vaccination against smallpox.

"Confirmatory bias" is the unconscious tendency to accept reports which support the reviewer's views and to downplay those which do not. Experimental studies show the problem exists in peer reviewing.

Publishing Process

The process of academic publishing, which begins when authors submit a manuscript to a publisher, is divided into two distinct phases: peer review and production.

The process of peer review is organized by the journal editor and is complete when the content of the article, together with any associated images or figures, are accepted for publication. The peer review process is increasingly managed online, through the use of proprietary systems, commercial software packages, or open source and free software. A manuscript undergoes one or more rounds of review; after each round, the author(s) of the article modify their submission in line with the reviewers' comments; this process is repeated until the editor is satisfied and the work is accepted.

The production process, controlled by a production editor or publisher, then takes an article through copy editing, typesetting, inclusion in a specific issue of a journal, and then printing and online publication. Academic copy editing seeks to ensure that an article conforms to the journal's house style, that all of the referencing and labelling is correct, and that the text is consistent and legible; often this work involves substantive editing and negotiating with the authors. Because the work of academic copy editors can overlap with that of authors' editors, editors employed by journal publishers often refer to themselves as "manuscript editors".

In much of the 20th century, such articles were photographed for printing into proceedings and journals, and this stage were known as *camera-ready* copy. With modern digital submission in formats such as PDF, this photographing step is no longer necessary, though the term is still sometimes used.

The author will review and correct proofs at one or more stages in the production process. The proof correction cycle has historically been labour-intensive as handwritten comments by authors and editors are manually transcribed by a proof reader onto a clean version of the proof. In the early 21st century, this process was streamlined by the introduction of e-annotations in Microsoft Word, Adobe Acrobat, and other programs, but it still remained a time-consuming and error-prone process. The full automation of the proof correction cycles has only become possible with the onset of online collaborative writing platforms, such as Authorea, Google Docs, and various others, where a remote service oversees the copy-editing interactions of multiple authors and exposes them as explicit, actionable historic events.

Citations

Academic authors cite sources they have used, in order to support their assertions and arguments and to help readers find more information on the subject. It also gives credit to authors whose work they use and helps avoid plagiarism.

Each scholarly journal uses a specific format for citations (also known as references). Among the most common formats used in research papers are the APA, CMS, and MLA styles.

The American Psychological Association (APA) style is often used in the social sciences. The Chicago Manual of Style (CMS) is used in business, communications, economics, and social sciences. The CMS style uses footnotes at the bottom of page to help readers locate the sources. The Modern Language Association (MLA) style is widely used in the humanities.

Publishing by Discipline

Natural Sciences

Scientific, technical, and medical (STM) literature is a large industry which generated $23.5 billion in revenue; $9.4 billion of that was specifically from the publication of English language scholarly journals. Most scientific research is initially published in scientific journals and considered to be a primary source. Technical reports, for minor research results and engineering and design work (including computer software) round out the primary literature. Secondary sources in the sciences include articles in review journals (which provide a synthesis of research articles on a topic to highlight advances and new lines of research), and books for large projects, broad arguments, or compilations of articles. Tertiary sources might include encyclopedias and similar works intended for broad public consumption or academic libraries.

A partial exception to scientific publication practices is in many fields of applied science, particularly that of U.S. computer science research. An equally prestigious site of publication within U.S. computer science are some academic conferences. Reasons for this departure include a large number of such conferences, the quick pace of research progress, and computer science professional society support for the distribution and archiving of conference proceedings.

Social Sciences

Publishing in the social sciences is very different in different fields. Some fields, like economics, may have very "hard" or highly quantitative standards for publication, much like the natural sciences. Others, like anthropology or sociology, emphasize field work and reporting on first-hand observation as well as quantitative work. Some social science fields, such as public health or demography, have significant shared interests with professions like law and medicine, and scholars in these fields often also publish in professional magazines.

Humanities

Publishing in the humanities is in principle similar to publishing elsewhere in the academy; a range of journals, from general to extremely specialized, are available, and university presses issue many new humanities books every year. The arrival of online publishing opportunities has radically transformed the economics of the field and the shape of the future is controversial. Unlike science, where timeliness is critically important, humanities publications often take years to write and years more to publish. Unlike the sciences, research is most often an individual process and is seldom supported by large grants. Journals rarely make profits and are typically run by university departments.

The following describes the situation in the United States. In many fields, such as literature and history, several published articles are typically required for a first tenure-track job, and a published or forthcoming *book* is now often required before tenure. Some critics complain that this *de facto* system has emerged without thought to its consequences; they claim that the predictable result is the publication of much shoddy work, as well as unreasonable demands on the already limited research time of young scholars. To make matters worse, the circulation of many humanities journals in the 1990s declined to almost untenable levels, as many libraries cancelled subscriptions, leaving fewer and fewer peer-reviewed outlets for publication; and many humanities professors' first books sell only a few hundred copies, which often does not pay for the cost of their printing. Some scholars have called for a publication subvention of a few thousand dollars to be associated with each graduate student fellowship or new tenure-track hire, in order to alleviate the financial pressure on journals.

Open Access Journals

An alternative to the subscription model of journal publishing is the open access journal model, which typically involves a publication charge being paid by the author. Prestige journals typically charge several thousand dollars. Oxford University Press, with over 300 journals, has fees ranging from £1000-£2500, with discounts of 50% to 100% to authors from developing countries. Wiley Blackwell has 700 journals available, and they charge a flat $US3000 open access fee. Springer, with over 2600 journals, charges US$3000 or EUR 2200 (excluding VAT).

The online distribution of individual articles and academic journals then takes place without charge to readers and libraries. Most open access journals remove all the financial, technical, and legal barriers that limit access to academic materials to paying customers. The Public Library of Science and BioMed Central are prominent examples of this model.

Open access has been criticized on quality grounds, as the desire to maximize publishing fees could cause some journals to relax the standard of peer review. It may be criticized on financial grounds as well, because the necessary publication fees have proven to be higher than originally expected. Open access advocates generally reply that because open access is as much based on peer reviewing as traditional publishing, the quality should be the same (recognizing that both traditional and open access journals have a range of quality). It has also been argued that good science done by academic institutions who cannot afford to pay for open access might not get published at all, but most open access journals permit the waiver of the fee for financial hardship or authors in underdeveloped countries. In any case, all authors have the option of self-archiving their articles in their institutional repositories in order to make them open access, whether or not they publish them in a journal.

If they publish in a Hybrid open access journal, authors pay a subscription journal a publication fee to make their individual article open access. The other articles in such hybrid journals are either made available after a delay, or remain available only by subscription. Most traditional publishers (including Wiley-Blackwell, Oxford University Press, and Springer Science+Business Media) have already introduced such a hybrid option, and more are following. Proponents of open access suggest that such moves by corporate publishers illustrate that open access, or a mix of open access and traditional publishing, can be financially viable, and evidence to that effect is emerging. The fraction of the authors of a hybrid open access journal that make use of its open access option can however be small. It also remains unclear whether this is practical in fields outside the sciences, where there is much less availability of outside funding. In 2006, several funding agencies, including the Wellcome Trust and several divisions of the Research Councils in the UK announced the availability of extra funding to their grantees for such open access journal publication fees.

In May 2016, the Council for the European Union agreed that from 2020 all scientific publications as a result of publicly funded research must be freely available. It also must be able to optimally reuse research data. To achieve that, the data must be made accessible, unless there are well-founded reasons for not doing so, for example intellectual property rights or security or privacy issues.

Academic Publishing Growth

In recent decades there has been a growth in academic publishing in developing countries as they become more advanced in science and technology. Although the large majority of scientific output and academic documents are produced in developed countries, the rate of growth in these countries has stabilized and is much smaller than the growth rate in some of the developing countries. The fastest scientific output growth rate over the last two decades has been in the Middle East and Asia with Iran leading with an 11-fold increase followed by the Republic of Korea, Turkey, Cyprus, China, and Oman. In comparison, the only G8 countries in top 20 ranking with fastest performance improvement are, Italy which stands at tenth and Canada at 13th globally.

By 2004, it was noted that the output of scientific papers originating from the European Union had a larger share of the world's total from 36.6 to 39.3 percent and from 32.8 to 37.5 per cent of the "top one per cent of highly cited scientific papers". However, the United States' output dropped 52.3 to 49.4 per cent of the world's total, and its portion of the top one percent dropped from 65.6 to 62.8 per cent.

Iran, China, India, Brazil, and South Africa were the only developing countries among the 31 nations that produced 97.5% of the most cited scientific articles in a study published in 2004. The remaining 162 countries contributed less than 2.5%. The Royal Society in a 2011 report stated that in share of English scientific research papers the United States was first followed by China, the UK, Germany, Japan, France, and Canada. The report predicted that China would overtake the United States some time before 2020, possibly as early as 2013. China's scientific impact, as measured by other scientists citing the published papers the next year, is smaller although also increasing.

Peer Review

A reviewer at the American National Institutes of Health evaluates a grant proposal.

Peer review is the evaluation of work by one or more people of similar competence to the producers of the work (peers). It constitutes a form of self-regulation by qualified members of a profession within the relevant field. Peer review methods are employed to maintain standards of quality, improve performance, and provide credibility. In academia, scholarly peer review is often used to determine an academic paper's suitability for publication. Peer review can be categorized by the type of activity and by the field or profession in which the activity occurs, e.g., medical peer review.

Professional

Professional peer review focuses on the performance of professionals, with a view to improving quality, upholding standards, or providing certification. In academia, peer review is common in decisions related to faculty advancement and tenure.

A prototype professional peer-review process was recommended in the *Ethics of the Physician* written by Ishāq ibn 'Alī al-Ruhāwī (854–931). He stated that a visiting physician had to make duplicate notes of a patient's condition on every visit. When the patient was cured or had died, the notes of the physician were examined by a local medical council of other physicians, who would decide whether the treatment had met the required standards of medical care.

Professional peer review is common in the field of health care, where it is usually called *clinical peer review*. Further, since peer review activity is commonly segmented by clinical discipline, there is also physician peer review, nursing peer review, dentistry peer review, etc. Many other professional fields have some level of peer review process: accounting, law, engineering (e.g., software peer review, technical peer review), aviation, and even forest fire management.

Peer review is used in education to achieve certain learning objectives, particularly as a tool to reach higher order processes in the affective and cognitive domains as defined by Bloom's taxonomy. This may take a variety of forms, including closely mimicking the scholarly peer review processes used in science and medicine.

Scholarly

Scholarly peer review (also known as refereeing) is the process of subjecting an author's scholarly work, research, or ideas to the scrutiny of others who are experts in the same field, before a paper describing this work is published in a journal or as a book. The peer review helps the publisher (that is, the editor-in-chief or the editorial board) decide whether the work should be accepted, considered acceptable with revisions, or rejected. Peer review requires a community of experts in a given (and often narrowly defined) field, who are qualified and able to perform reasonably impartial review. Impartial review, especially of work in less narrowly defined or inter-disciplinary fields, may be difficult to accomplish, and the significance (good or bad) of an idea may never be widely appreciated among its contemporaries. Peer review is generally considered necessary to academic quality and is used in most major scientific journals, but does by no means prevent publication of all invalid research. Traditionally, peer reviewers have been anonymous, but there is currently a significant amount of *open peer review*, where the comments are visible to readers, generally with the identities of the peer reviewers disclosed as well.

Government Policy

The European Union has been using peer review in the 'Open Method of Co-ordination' of policies in the fields of active labour market policy since 1999. In 2004, a program of peer reviews started in social inclusion. Each program sponsors about eight peer review meetings in each year, in which a 'host country' lays a given policy or initiative open to examination by half a dozen other countries and the relevant European-level NGOs. These usually meet over two days and include visits to local sites where the policy can be seen in operation. The meeting is preceded by the compilation of an expert report on which participating 'peer countries' submit comments. The results are published on the web.

The United Nations Economic Commission for Europe, through UNECE Environmental Performance Reviews, uses the technique of peer review to evaluate progress made by its member countries in improving their environmental policies.

The State of California is the only U.S. state to mandate scientific peer review. In 1997, the California Governor signed into law Senate Bill 1320 (Sher), Chapter 295, statutes of 1997, which mandates that, before any CalEPA Board, Department, or Office adopts a final version of a rule-making, the scientific findings, conclusions, and assumptions on which the proposed rule are based must be submitted for independent external scientific peer review. This requirement is incorporated into the California Health and Safety Code Section 57004.

Medical

Medical peer review may refer to clinical peer review, or the peer evaluation of clinical teaching skills for both physicians and nurses, or scientific peer review of journal articles, or to a secondary round of peer review for the clinical value of articles concurrently published in medical journals. "Medical peer review" has been used by the American Medical Association to refer not only to the process of improving quality and safety in health care organizations, but also to the process of rating clinical behavior or compliance with professional society membership standards. Thus, the terminology has poor standardization and specificity, particularly as a database search term.

Process, Costs and Criticisms

It has been argued that peer review is impossible to define in operational terms; someone performing exactly the same research could just as easily be a financial competitor as a collaborator, creating potential for conflict of interest. Definitions of what constitute the review process may vary across journals and disciplines. A study by the BMJ inserted deliberate errors into publication which successfully avoided fact checking.

Journals may take over a year to publish. Most reviewers are not paid, with the cost of review for the BMJ estimated at £100 per paper. On average the academic community pays roughly $5000 for access to a peer reviewed paper.

There is strong evidence of bias against women in the process of awarding grants. The editorial peer review process has also been strongly biased against `negative studies', those findings where in intervention does not work.

One randomized trial found blinding reviewers to the identity of authors improved the quality of reviews), although this presented a means of experimentally assessing peer review, two later studies found contrary findings that blinding reviewers improved the quality of reviews. These studies also showed that such blinding is difficult to achieve, and that reviewers could identify the authors in about a quarter to a third of cases.

Scholarly Peer Review

Scholarly peer review (also known as refereeing) is the process of subjecting an author's scholarly work, research, or ideas to the scrutiny of others who are experts in the same field, before a paper describing this work is published in a journal or as a book. The peer review helps the publisher (that is, the editor-in-chief or the editorial board) decide whether the work should be accepted, considered acceptable with revisions, or rejected. Peer review requires a community of experts in a given (and often narrowly defined) field, who are qualified and able to perform reasonably impartial review. Impartial review, especially of work in less narrowly defined or inter-disciplinary fields, may be difficult to accomplish, and the significance (good or bad) of an idea may never be widely appreciated among its contemporaries. Peer review is generally considered necessary to academic quality and is used in most major scientific journals, but does by no means prevent publication of all invalid research. Traditionally, peer reviewers have been anonymous, but there is currently a significant amount of *open peer review*, where the comments are visible to readers, generally with the identities of the peer reviewers disclosed as well.

History

The first record of an editorial pre-publication peer-review is from 1665 by Henry Oldenburg, the founding editor of *Philosophical Transactions of the Royal Society* at the Royal Society of London.

The first peer-reviewed publication might have been the *Medical Essays and Observations* published by the Royal Society of Edinburgh in 1731. The present-day peer-review system evolved from this 18th-century process, began to involve external reviewers in the mid-19th-century, and did not become commonplace until the mid-20th-century.

Peer review became a touchstone of the scientific method, but until the end of the 19th century only an editor-in-chief or editorial committees performed it.

Editors of scientific journals made publication decisions without seeking outside input, i.e. an external panel of reviewers, giving established authors latitude in their journalistic discretion. For example, Albert Einstein's four revolutionary *Annus Mirabilis* papers in the 1905 issue of *Annalen der Physik* were peer-reviewed by the journal's editor-in-chief, Max Planck, and its co-editor, Wilhelm Wien, both future Nobel prize winners and together experts on the topics of these papers. On another occasion, Einstein was severely critical of the external review process, saying that he had not authorized the editor in chief to show his manuscript "to specialists before it is printed", and informing him that he would "publish the paper elsewhere". While some medical journals started to systematically appoint external reviewers, it is only since the middle of the 20th century that this practice has spread widely and that external reviewers have been given some visibility within academic journals, including being thanked by authors and editors. A 2003 editorial in *Nature* stated that "in journals in those days, the burden of proof was generally on the opponents rather than the proponents of new ideas.". The journal Nature itself instituted formal peer review only in 1967.

In the 20th century, peer review also became common for science funding allocations. This process appears to have developed independently from that of editorial peer review. Gaudet, provides a social science view of the history of peer review carefully tending to what is under investigation, here peer review, and not only looking at superficial or self-evident commonalities among inquisition, censorship, and journal peer review. It builds on historical research by Gould, Biagioli, Spier, and Rip. The first Peer Review Congress met in 1989. Over time, the fraction of papers devoted to peer review has steadily declined, suggesting that as a field of sociological study, it has been replaced by more systematic studies of bias and errors.

In parallel with 'common experience' definitions based on the study of peer review as a 'pre-constructed process', some social scientists have looked at peer review without considering it as pre-constructed. Hirschauer proposed that journal peer review can be understood as reciprocal accountability of judgements among peers. Gaudet proposed that journal peer review could be understood as a social form of boundary judgement - determining what can be considered as scientific (or not) set against an overarching knowledge system, and following predecessor forms of inquisition and censorship.

Pragmatically, peer review refers to the work done during the screening of submitted manuscripts. This process encourages authors to meet the accepted standards of their discipline and reduces the dissemination of irrelevant findings, unwarranted claims, unacceptable interpretations, and personal views. Publications that have not undergone peer review are likely to be regarded with suspicion by academic scholars and professionals. Non-peer-reviewed work does not contribute, or contributes less, to the academic credit of scholar such as the h-index, although this heavily depends on the field.

Justification

It is difficult for authors and researchers, whether individually or in a team, to spot every mistake or flaw in a complicated piece of work. This is not necessarily a reflection on those concerned, but because with a new and perhaps eclectic subject, an opportunity for improvement may be more obvious to someone with special expertise or who simply looks at it with a fresh eye. Therefore, showing work to others increases the probability that weaknesses will be identified and improved. For both grant-funding and publication in a scholarly journal, it is also normally a requirement that the subject is both novel and substantial.

The decision whether or not to publish a scholarly article, or what should be modified before publication, ultimately lies with the publisher (editor-in-chief or the editorial board) to which the manuscript has been submitted. Similarly, the decision whether or not to fund a proposed project rests with an official of the funding agency. These individuals usually refer to the opinion of one or more reviewers in making their decision. This is primarily for three reasons:

- Workload. A small group of editors/assessors cannot devote sufficient time to each of the many articles submitted to many journals.
- Miscellany of ideas. Were the editor/assessor to judge all submitted material themselves, approved material would solely reflect their opinion.
- Limited expertise. An editor/assessor cannot be expected to be sufficiently expert in all areas covered by a single journal or funding agency to adequately judge all submitted material.

Reviewers are often anonymous and independent. However, some reviewers may choose to waive their anonymity, and in other limited circumstances, such as the examination of a formal complaint against the referee, or a court order, the reviewer's identity may have to be disclosed. Anonymity may be unilateral or reciprocal (single- or double-blinded reviewing).

Since reviewers are normally selected from experts in the fields discussed in the article, the process of peer review helps to keep some invalid or unsubstantiated claims out of the body of published research and knowledge. Scholars will read published articles outside their limited area of detailed expertise, and then rely, to some degree, on the peer-review process to have provided reliable and credible research that they can build upon for subsequent or related research. Significant scandal ensues when an author is found to have falsified the research included in an article, as other scholars, and the field of study itself, may have relied upon the invalid research.

For US universities, peer reviewing of books before publication is a requirement for full membership of the Association of American University Presses.

Procedure

In the case of proposed publications, the publisher (editor-in-chief or the editorial board, often with assistance of corresponding ediors) sends advance copies of an author's work or ideas to researchers or scholars who are experts in the field (known as "referees" or "reviewers"), nowadays normally by e-mail or through a web-based manuscript processing system. Depending on the field of study and on the specific journal, there are usually one to three referees for a given article.

These referees each return an evaluation of the work to the publisher, noting weaknesses or problems along with suggestions for improvement. Typically, most of the referees' comments are eventually seen by the author, though a referee can also send 'for your eyes only' comments to the publisher; scientific journals observe this convention almost universally. The publisher, usually familiar with the field of the manuscript (although typically not in as much depth as the referees, who are specialists), then evaluates the referees' comments, her or his own opinion of the manuscript, and the context of the scope of the journal or level of the book and readership, before passing a decision back to the author(s), usually with the referees' comments.

Referees' evaluations usually include an explicit recommendation of what to do with the manuscript or proposal, often chosen from options provided by the journal or funding agency. Most recommendations are along the lines of the following:

- to unconditionally accept the manuscript or the proposal,
- to accept it in the event that its authors improve it in certain ways,
- to reject it, but encourage revision and invite resubmission,
- to reject it outright.

During this process, the role of the referees is advisory. The publisher is typically under no obligation to accept the opinions of the referees, though she will most often do so. Furthermore, the referees in scientific publication do not act as a group, do not communicate with each other, and typically are not aware of each other's identities or evaluations. Proponents argue that if the reviewers of a paper are unknown to each other, the publisher can more easily verify the objectivity of the reviews. There is usually no requirement that the referees achieve consensus, with the decision instead often made by the publisher based on her best judgement of the arguments. The group dynamics are thus substantially different from that of a jury.

In situations where multiple referees disagree substantially about the quality of a work, there are a number of strategies for reaching a decision. When a publisher receives very positive and very negative reviews for the same manuscript, the publisher will often solicit one or more additional reviews as a tie-breaker. As another strategy in the case of ties, the publisher may invite authors to reply to a referee's criticisms and permit a compelling rebuttal to break the tie. If a publisher does not feel confident to weigh the persuasiveness of a rebuttal, the publisher may solicit a response from the referee who made the original criticism. A publisher may convey communications back and forth between authors and a referee, in effect allowing them to debate a point. Even in these cases, however, publishers do not allow multiple referees to confer with each other, though each reviewer may often see earlier comments submitted by other reviewers. The goal of the process is explicitly not to reach consensus or to persuade anyone to change their opinions, but instead to provide material for an informed editorial decision. Some medical journals, usually following the open access model, have begun posting on the Internet the pre-publication history of each individual article, from the original submission to reviewers' reports, authors' comments, and revised manuscripts.

Traditionally, reviewers would often remain anonymous to the authors, but this standard varies both with time and with academic field. In some academic fields, most journals offer the reviewer the option of remaining anonymous or not, or a referee may opt to sign a review, thereby relinquishing anonymity. Published papers sometimes contain, in the acknowledgments section, thanks to anonymous or named referees who helped improve the paper.

In some disciplines there exist refereed venues (such as conferences and workshops). To be admitted to speak, scholars and scientists must submit papers (generally short, often 15 pages or less) in advance. These papers are reviewed by a "program committee" (the equivalent of an editorial board), which generally requests inputs from referees. The hard deadlines set by the conferences tend to limit the options to either accepting or rejecting the paper.

Recruiting Referees

At a journal or book publisher, the task of picking reviewers typically falls to an editor. When a manuscript arrives, an editor solicits reviews from scholars or other experts who may or may not have already expressed a willingness to referee for that journal or book division. Granting agencies typically recruit a panel or committee of reviewers in advance of the arrival of applications.

Referees are supposed to inform the editor of any conflict of interests that might arise. Journals or individual editors may invite a manuscript's authors to name people whom they consider qualified to referee their work. For some journals this is a requirement of submission. Authors are sometimes also given the opportunity to name natural candidates who should be *disqualified*, in which case they may be asked to provide justification (typically expressed in terms of conflict of interest).

Editors solicit author input in selecting referees because academic writing typically is very specialized. Editors often oversee many specialties, and can not be experts in all of them. But after an editor selects referees from the pool of candidates, the editor typically is obliged not to disclose the referees' identities to the authors, and in scientific journals, to each other. Policies on such matters differ among academic disciplines. One difficulty with respect to some manuscripts is that, there may be few scholars who truly qualify as experts, people who have themselves done work similar to that under review. This can frustrate the goals of reviewer anonymity and avoidance of conflicts of interest. Low-prestige or local journals and granting agencies that award little money are especially handicapped with regard to recruiting experts.

A potential hindrance in recruiting referees is that they are usually not paid, largely because doing so would itself create a conflict of interest. Also, reviewing takes time away from their main activities, such as his or her own research. To the would-be recruiter's advantage, most potential referees are authors themselves, or at least readers, who know that the publication system requires that experts donate their time. Serving as a referee can even be a condition of a grant, or professional association membership.

Referees have the opportunity to prevent work that does not meet the standards of the field from being published, which is a position of some responsibility. Editors are at a special advantage in recruiting a scholar when they have overseen the publication of his or her work, or if the scholar is one who hopes to submit manuscripts to that editor's publishing entity in the future. Granting agencies, similarly, tend to seek referees among their present or former grantees.

Peerage of Science is an independent service and a community where reviewer recruitment happens via Open Engagement: authors submit their manuscript to the service where it is made accessible for any non-affiliated scientist, and 'validated users' choose themselves what they want to review. The motivation to participate as a peer reviewer comes from a reputation system where the quality of the reviewing work is judged and scored by other users, and contributes to user profiles. Peerage of Science does not charge any fees to scientists, and does not pay peer reviewers. Participating publishers however pay to use the service, gaining access to all ongoing processes and the opportunity to make publishing offers to the authors.

With independent peer review services the author usually retains the right to the work throughout the peer review process, and may choose the most appropriate journal to submit the work to. Peer review services may also provide advice or recommendations on most suitable journals for the work. Journals may still want to perform an independent peer review, without the potential conflict of interest that financial reimbursement may cause, or the risk that an author has contracted multiple peer review services but only presents the most favorable one.

An alternative or complementary system of performing peer review is for the author to pay for having it performed. Example of such service provider is *Rubriq*, which for each work assigns peer reviewers who are financially compensated for their efforts.

Different Styles

Anonymous and Attributed

For most scholarly publications, the identity of the reviewers is kept anonymised (also called "blind peer review). The alternative, *attributed peer review* involves revealing the identities of the reviewers. Some reviewers choose to waive their right to anonymity, even when the journal's default format is blind peer review.

In anonymous peer review, reviewers are known to the journal editor or conference organiser but their names are not given to the article's author. In some cases, the author's identity can also be anonymised for the review process, with identifying information is stripped from the document before review. The system is intended to reduce or eliminate bias.

Others support blind reviewing because no research has suggested that the methodology may be harmful and that the cost of facilitating such reviews is minimal. Some experts proposed blind review procedures for reviewing controversial research topics.

In "double-blind" review, which has been fashioned by sociology journals in the 1950s and remains more common in the social sciences and humanities than in the natural sciences, the identity of the authors is concealed from the reviewers, and vice versa, lest the knowledge of authorship or concern about disapprobation from the author bias their review. Critics of the double-blind review process point out that, despite any editorial effort to ensure anonymity, the process often fails to do so, since certain approaches, methods, writing styles, notations, etc., point to a certain group of people in a research stream, and even to a particular person.

In many fields of "big science", the publicly available operation schedules of major equipments, such as telescopes or synchrotrons, would make the authors' names obvious to anyone who would care to look them up. Proponents of double-blind review argue that it performs no worse than single-blind, and that it generates a perception of fairness and equality in academic funding and publishing. Single-blind review is strongly dependent upon the goodwill of the participants, but no more so than double-blind review with easily identified authors.

As an alternative to single-blind and double-blind review, authors and reviewers are encouraged to declare their conflicts of interest when the names of authors and sometimes reviewers are known to the other. When conflicts are reported, the conflicting reviewer can be prohibited from reviewing and discussing the manuscript, or his or her review can instead be interpreted with the reported conflict in mind; the latter option is more often adopted when the conflict of interest is mild, such as a previous professional connection or a distant family relation. The incentive for reviewers to declare their conflicts of interest is a matter of professional ethics and individual integrity. Even when the reviews are not public, they are still a matter of record and the reviewer's credibility depends upon how they represent themselves among their peers. Some software engineering journals, such as the *IEEE Transactions on Software Engineering*, use non-blind reviews with reporting to editors of conflicts of interest by both authors and reviewers.

A more rigorous standard of accountability is known as an audit. Because reviewers are not paid, they cannot be expected to put as much time and effort into a review as an audit requires. Therefore, academic journals such as *Science*, organizations such as the American Geophysical Union, and agencies such as the National Institutes of Health and the National Science Foundation maintain and archive scientific data and methods in the event another researcher wishes to replicate or audit the research after publication.

The traditional anonymous peer review has been criticized for its lack of accountability, the possibility of abuse by reviewers or by those who manage the peer review process, its possible bias, and its inconsistency, alongside other flaws. Eugene Koonin, a senior investigator at the National Center for Biotechnology Information, asserts that the system has "well-known ills"and advocates "open peer review".

Open Peer Review

Starting in the 1990s, several scientific journals started experiments with hybrid peer review processes, allowing the open peer reviews in parallel to the traditional model. The initial evidence of the effects of open peer reviews was mixed. Identifying reviewers to the authors does not negatively impact, and may potentially have a positive impact upon, the quality of reviews, the recommendation regarding publication, the tone of the review and the time spent on reviewing. However, more of those who are invited to review decline to do so. Informing reviewers that their signed reviews might be posted on the web and available to the wider public did not have a negative impact on quality of reviews and recommendations regarding publication, but it led to a longer time spent on reviewing, besides a higher reviewer decline rate. The results suggest that open peer review is feasible, and does not lead to poorer quality of reviews, but needs to be balanced against the increase in review time, and higher decline rates among invited reviewers.

A number of reputable medical publishers have trialed the open peer review concept. The first open peer review trial was conducted by The Medical Journal of Australia (MJA) in cooperation with the University of Sydney Library, from March 1996 to June 1997. In that study 56 research articles accepted for publication in the MJA were published online together with the peer reviewers' comments; readers could email their comments and the authors could amend their articles further before print publication of the article. The investigators concluded that the process had modest benefits for authors, editors and readers.

Allegations of Bias and Suppression

The interposition of editors and reviewers between authors and readers may enable the intermediators to act as gatekeepers. Some sociologists of science argue that peer review makes the ability to publish susceptible to control by elites and to personal jealousy. The peer review process may suppress dissent against "mainstream" theories and may be biased against novelty. Reviewers tend to be especially critical of conclusions that contradict their own views, and lenient towards those that match them. At the same time, established scientists are more likely than others to be sought out as referees, particularly by high-prestige journals/publishers. As a result, ideas that harmonize with the established experts' are more likely to see print and to appear in premier journals than are iconoclastic or revolutionary ones. This accords with Thomas Kuhn's well-known observations regarding scientific revolutions. A theoretical model has been established whose simulations imply that peer review and over-competitive research funding foster mainstream opinion to monopoly.

Criticisms of traditional anonymous peer review allege that it lacks accountability, can lead to abuse by reviewers, and may be biased and inconsistent.

Failures

Peer review fails when a peer-reviewed article contains fundamental errors that undermine at least one of its main conclusions and that could have been identified by more careful reviewers. Many journals have no procedure to deal with peer review failures beyond publishing letters to the editor.

Peer review in scientific journals assumes that the article reviewed has been honestly prepared. The process occasionally detects fraud, but is not designed to do so. When peer review fails and a paper is published with fraudulent or otherwise irreproducible data, the paper may be retracted.

A 1998 experiment on peer review with a fictitious manuscript found that peer reviewers failed to detect some manuscript errors and the majority of reviewers may not notice that the conclusions of the paper are unsupported by its results.

Fake

There have been instances where peer review was claimed to be performed but in fact was not; this has been documented in some predatory open access journals (e.g., the *Who's Afraid of Peer Review?* affair) or in the case of sponsored Elsevier journals.

In November 2014, an article in *Nature* exposed that some academics were submitting fake contact details for recommended reviewers to journals, so that if the publisher contacted the recommended reviewer, they were the original author reviewing their own work under a fake name. The Committee on Publication Ethics issued a statement warning of the fraudulent practice. In March 2015, Biomed Central retracted 43 articles and Springer retracted 64 papers in 10 journals in August 2015.

Plagiarism

Reviewers generally lack access to raw data, but do see the full text of the manuscript, and are typically familiar with recent publications in the area. Thus, they are in a better position to detect plagiarism of prose than fraudulent data. A few cases of such textual plagiarism by historians, for instance, have been widely publicized.

On the scientific side, a poll of 3,247 scientists funded by the U.S. National Institutes of Health found 0.3% admitted faking data and 1.4% admitted plagiarism. Additionally, 4.7% of the same poll admitted to self-plagiarism or autoplagiarism, in which an author republishes the same material, data, or text, without citing their earlier work.,

Abuse of Inside Information by Reviewers

A related form of professional misconduct is a reviewer using the not-yet-published information from a manuscript or grant application for personal or professional gain. The frequency with which this happens is unknown, but the United States Office of Research Integrity has sanctioned reviewers who have been caught exploiting knowledge they gained as reviewers. A possible defense for authors against this form of misconduct on the part of reviewers is to pre-publish their work in the form of a preprint or technical report on a public system such as arXiv. The preprint can later be used to establish priority, although preprints violate the stated policies of some journals.

Examples

- "Perhaps the most widely recognized failure of peer review is its inability to ensure the identification of high-quality work. The list of important scientific papers that were rejected by some peer-reviewed journals goes back at least as far as the editor of *Philosophical Transaction's* 1796 rejection of Edward Jenner's report of the first vaccination against smallpox."
- The Soon and Baliunas controversy involved the publication in 2003 of a review study written by aerospace engineer Willie Soon and astronomer Sallie Baliunas in the journal *Climate Research*, which was quickly taken up by the G.W. Bush administration as a basis for amending the first Environmental Protection Agency *Report on the Environment*. The paper was strongly criticized by numerous scientists for its methodology and for its misuse of data from previously published studies, prompting concerns about the peer review process of the paper. The controversy resulted in the resignation of several editors of the journal and the admission by its publisher Otto Kinne that the paper should not have been published as it was.

- The trapezoidal rule, in which the method of Riemann sums for numerical integration was republished in a Diabetes research journal, *Diabetes Care*. The method is almost always taught in high school calculus, and was thus considered an example of an extremely well known idea being re-branded as a new discovery.
- A conference organized by the Wessex Institute of Technology was the target of an exposé by three researchers who wrote nonsensical papers (including one that was composed of random phrases). They reported that the papers were "reviewed and provisionally accepted" and concluded that the conference was an attempt to "sell" publication possibilities to less experienced or naive researchers. This may however be better described as a lack of any actual peer review, rather than peer review having failed.
- In 2014, an editorial was published in *Nature* highlighting problems with the peer-review process.

Improvement Efforts

Efforts to make fundamental improvements have ebbed and flowed since the late 1970s when Rennie first systematically reviewed articles in thirty medical journals. According to Ana Marušić, "Nothing much has changed in 25 years". Mentorship has not been shown to have a positive effect. Worse, little evidence indicates that peer review as presently performed, improves the quality of published papers.

An extension of peer review beyond the date of publication is *open peer commentary*, whereby expert commentaries are solicited on published articles and the authors are encouraged to respond. It was first implemented by the anthropologist Sol Tax, who founded the journal Current Anthropology, published by University of Chicago Press in 1959. The journal *Behavioral and Brain Sciences*, published by Cambridge University Press, was founded by Stevan Harnad in 1978 and modeled on Current Anthropology's open peer commentary feature. Psycoloquy was founded in 1990 on the basis of the same feature, but this time implemented online.

In the summer of 2009, Kathleen Fitzpatrick explored open peer review and commentary in her book, *Planned Obsolescence*. Throughout the 2000s academic journals based solely on the concept of open peer review were launched, such as *Philica*.

Early Era: 1996–2000

In 1996, the *Journal of Interactive Media in Education* launched using open peer review. Reviewers' names were made public, they were therefore accountable for their review, and their contribution was acknowledged. Authors had the right of reply, and other researchers had the chance to comment prior to publication. As of February 2013, the "Journal of Interactive Media in Education" stopped using open peer review.

In 1997, the *Electronic Transactions on Artificial Intelligence* was launched as an open access journal by the European Coordinating Committee for Artificial Intelligence. This journal used a two-stage review process. In the first stage, papers that passed a quick screen by the editors were immediately published on the Transaction's discussion website for the purpose of on-line public discussion during a period of at least three months, where the contributors' names were made public except in exceptional cases. At the end of the discussion period, the authors were invited to submit a revised version of the article, and anonymous referees decided whether the revised manuscript would be accepted to the journal or not, but without any option for the referees to propose further changes. The last issue of this journal appeared in 2001.

In 1999, the open access journal *Journal of Medical Internet Research* was launched, which from its inception decided to publish the names of the reviewers at the bottom of each published article. Also in 1999, the *British Medical Journal* moved to an open peer review system, revealing reviewers' identities to the authors but not the readers, and in 2000, the medical journals in the open access BMC series published by BioMed Central, launched using open peer review. As with the *BMJ*, the reviewers' names are included on the peer review reports. In addition, if the article is published the reports are made available online as part of the 'pre-publication history'.

Several other journals published by the BMJ Group allow optional open peer review, as does *PLoS Medicine*, published by the Public Library of Science. The *BMJ*'s Rapid Responses allows ongoing debate and criticism following publication.

Recent Era: 2001–present

Atmospheric Chemistry and Physics (ACP), an open access journal launched in 2001 by the European Geosciences Union, has a two-stage publication process. In the first stage, papers that pass a quick screen by the editors are immediately published on the *Atmospheric Chemistry and Physics Discussions* (ACPD) website. They are then subject to interactive public discussion alongside formal peer review. Referees' comments (either anonymous or attributed), additional short comments by other members of the scientific community (which must be attributed) and the authors' replies are also published in *ACPD*. In the second stage, the peer-review process is completed and, if the article is formally accepted by the editors, the final revised papers are published in *ACP*. The success of this approach is shown by the ranking by Thomson Reuters of *ACP* as the top journal in the field of Meteorology & Atmospheric Sciences.

In June 2006, *Nature* launched an experiment in parallel open peer review: some articles that had been submitted to the regular anonymous process were also available online for open, identified public comment. The results were less than encouraging – only 5% of authors agreed to participate in the experiment, and only 54% of those articles received comments. The editors have suggested that researchers may have been too busy to take part and were reluctant to make their names public. The knowledge that articles were simultaneously being subjected to anonymous peer review may also have affected the uptake.

In February 2006, the journal *Biology Direct* was launched by BioMed Central, adding another alternative to the traditional model of peer review. If authors can find three members of the Editorial Board who will each return a report or will themselves solicit an external review, the article will be published. As with *Philica*, reviewers cannot suppress publication, but in contrast to *Philica*, no reviews are anonymous and no article is published without being reviewed. Authors have the opportunity to withdraw their article, to revise it in response to the reviews, or to publish it without revision. If the authors proceed with publication of their article despite critical comments, readers can clearly see any negative comments along with the names of the reviewers. In the social sciences, there have been experiments with wiki-style, signed peer reviews, for example in an issue of the *Shakespeare Quarterly.*

In 2010, the British Medical Journal began publishing signed reviewer's reports alongside accepted papers, after determining that telling reviewers that their signed reviews might be posted publicly did not significantly affect the quality of the reviews.

In 2011, Peerage of Science, and independent peer review service, was launched with several non-traditional approaches to academic peer review. Most prominently, these include the judging and scoring of the accuracy and justifiability of peer reviews, and concurrent usage of a single peer review round by several participating journals.

Starting in 2013 with the launch of *F1000Research*, some publishers have combined open peer review with postpublication peer review by using a versioned article system. At *F1000Research*, articles are published before review, and invited peer review reports (and reviewer names) are published with the article as they come in. Author-revised versions of the article are then linked to the original. A similar postpublication review system with versioned articles is used by *Science Open* and *The Winnower*, both launched in 2014.

In 2014, *Life* implanted an open peer review system, under which the peer-review reports and authors' responses are published as an integral part of the final version of each article.

Another form of "open peer review" is community-based pre-publication peer-review, where the review process is open for everybody to join.

Clinical Peer Review

Clinical Peer Review is the process by which health care professionals evaluate each other's clinical performance. Clinical peer review is segmented by discipline. No inter-disciplinary models for clinical peer review have been described. Physician Peer Review is most common and is found in virtually all hospitals. Peer review is also done in some settings by other clinical disciplines including nursing and pharmacy. Initially used by Dans, Clinical Peer Review is the best term to collectively refer to all such activity.

Medical peer review is the process by which a committee of physicians examines the work of a peer and determines whether the physician under review has met accepted standards of care in rendering medical services. Depending on the specific institution, a medical peer review may be initiated at the request of a patient, a physician, or an insurance carrier. The term "peer review" is sometimes used synonymously with performance appraisal.

Definitions

The definition of a peer review body can be broad, including not only individuals but also (for example, in Oregon), "tissue committees, governing bodies or committees including medical staff committees of a [licensed] health care facility...or any other medical group in connection with bona fide medical research, quality assurance, utilization review, credentialing, education, training, supervision or discipline of physicians or other health care providers." Definition of Peer Review

The first definition of nursing peer review was published in 1988 by the American Nurses Association and is still applicable today. This definition includes the following statements: "The American Nurses Association believes nurses bare primary responsibility and accountability for the quality of nursing care their clients receive. Standards of nursing practice provide a means for measuring the quality of nursing care a client receives. Each nurse is responsible for interpreting and implementing the standards of nursing practice. Likewise, each nurse must participate with other nurses in the decision-making process for evaluating nursing care...Peer review implies that the nursing care delivered by a group of nurses or an individual nurse is evaluated by individuals of the same rank or standing according to established standards of practice.... Peer review is an organized effort whereby practicing professionals review the quality and appropriateness of services ordered or performed by their professional peers. Peer review in nursing is the process by which practicing registered nurses systematically access, monitor, and make judgments about the quality of nursing care provided by peers as measured against professional standards of practice" (ANA 1988 p. 3).

Clinical peer review should be distinguished from the peer review that medical journals use to evaluate the merits of a scientific manuscript, from the peer review process used to evaluate health care research grant applications, and, also, from the process by which clinical teaching might be evaluated. All these forms of peer review are confounded in the term Medical peer review. Moreover, Medical peer review has been used by the American Medical Association (AMA) to refer not only to the process of improving quality and safety in health care organizations, but also to process by which adverse actions involving clinical privileges or professional society membership may be pursued.

History

The first documented description of a peer review process is found in the *Ethics of the Physician* written by Ishap bin Ali al-Rahawi (854–931) of al-Raha, Syria, who describes the first medical peer review process. His work, as well as later Arabic medical manuals, states that a visiting physician must always make duplicate notes of a patient's condition on every visit. When the patient was cured or had died, the notes of the physician were examined by a local medical council of other physicians, who would review the practising physician's notes to decide whether his or her performance met the required standards of medical care. If their reviews were negative, the practicing physician could face a lawsuit from a maltreated patient.

Medical audit, which remains the predominant mode of peer review in Europe, is a focused study of the process and/or outcomes of care for a specified patient cohort using pre-defined criteria, focused on a diagnosis, procedure or clinical situation. This audit process was revised by changes to The Joint Commission standards were revised in 1979, dispensing with the audit requirement and calling for an organized system of Quality Assurance (QA). Thus the objective of a medical peer review committee became, to investigate the medical care rendered in order to determine whether accepted standards of care have been met. Contemporaneous with this change, hospitals and physicians adopted generic screening to improve quality of care, despite warnings from the developers of these screens that they were not validated for this purpose, having originally been developed to evaluate no-fault malpractice insurance plans.

The focus on the question of whether the standard of care had been met persisted despite many criticisms, but is increasingly recognized to be outdated, replaced over the past decade by Quality Improvement (QI) principles.

Overview

The objective of a medical peer review committee is to investigate the medical care rendered in order to determine whether accepted standards of care have been met. The professional or personal conduct of a physician or other healthcare professional may also be investigated. If a medical peer review committee finds that a physician has departed from accepted standards, it may recommend limiting or terminating the physician's privileges at an institution. Remedial measures including education may also be recommended.

In Nursing, as in other professions, peer review applies professional control to practice, and is used by professionals to hold themselves accountable for their services to the public and the organization. Peer review plays a role in affecting the quality of outcomes, fostering practice development, and maintaining professional autonomy. The American Nurses Association guidelines on peer review define peer review as the process by which practitioners of the same rank, profession, or setting critically appraise each other's work performance against established standards. Professionals, who are best acquainted with the requirements and demands of the role, are the givers and receivers of the feedback review.

The medical peer review system is a quasi-judicial one, similar in some ways to the grand jury / petit jury system. First, a plaintiff asks for an investigation. Discretionary appointments of staff members are made by the medical Chief of Staff to create an *ad hoc committee*, which then conducts an investigation in the manner it feels is appropriate. There is no standard for due process, impartiality, or information sources; the review may consult the literature or an outside expert.

An indicted (and sanctioned) physician may have the right to request a hearing, with counsel allowed. A second panel of physicians is chosen as the 'petit jury', and a hearing officer is chosen. The accused physician has the option to demonstrate conflicts of interest and attempt to disqualify jurors based on reasonable suspicions of bias or conflicts of interest in a process akin to voir dire.

The Patient Safety and Quality Improvement Act of 2005 (Public Law 109-41) created Patient Safety Organizations, whose participants are immune from prosecution in civil, criminal, and administrative hearings, in order to act in parallel with peer review boards, using root cause analysis and evaluation of "near misses" in systems failure analysis.

Types

By Physicians

Today, physician peer review is most commonly done in hospitals, but may also occur in other practice settings including surgical centers and large group practices. The primary purpose of peer review is to improve the quality and safety of care. Secondarily, it serves to reduce the organization's vicarious malpractice liability and meet regulatory requirements. In the US, these include accreditation, licensure and Medicare participation. Peer review also supports the other processes that healthcare organizations have in place to assure that physicians are competent and practice within the boundaries of professionally accepted norms.

In varying degrees, physicians having been doing peer review for a long time. Peer review has been well documented in the 11th century and likely originated much earlier. In the 1900s, peer review methods appear to have evolved in relation to the pioneering work of Codman's End Result System and Ponton's concept of Medical Audit. Lembcke, himself a major contributor to audit methodology, in reviewing this history, noted the pre-emptive influence of hospital standardization promoted by the American College of Surgeons (ACS) following WWI. The Joint Commission on Accreditation of Hospitals followed the ACS in this role from 1952. Medicare legislation, enacted in 1964, was a boon to the Joint Commission. The conditions for hospital participation required a credible medical care review program. The regulations further stipulated that Joint Commission accreditation would guarantee payment eligibility. What was once a sporadic process, became hardwired in most hospitals following the Audit model. The widespread creation of new programs was hampered, however, by limitations in the available process models, tools, training and implementation support.

Medical audit is a focused study of the process and/or outcomes of care for a specified patient cohort using pre-defined criteria. Audits are typically organized around a diagnosis, procedure or clinical situation. The audit process can be effective in improving clinical performance. Clinical peer review remains the predominant mode of peer review in Europe.

In the 70s, the widespread creation of new programs was hampered by limitations in the available process models, tools, training and implementation support. The lack of perceived effectiveness of medical audit led to revisions of Joint Commission standards in 1979. Those modified standards dispensed with the audit requirement and called for an organized system of Quality Assurance (QA). About the same time, hospital and physicians were faced escalating malpractice insurance costs. In response to these combined pressures, they began to adopt "generic screens" for potential substandard care. These screens were originally developed to evaluate the feasibility of a no-fault medical malpractice insurance plan and were never validated as a tool to improve quality of care. Despite warnings from the developers, their use became widespread. In the process, a QA model for peer review evolved with a narrow focus on the question of whether or not the standard of care had been met. It has persisted despite the many criticisms of its methods and effectiveness. Today, its methods are increasingly recognized to be outdated and incongruent with the Quality Improvement (QI) principles that have been successfully adopted into the field of health care over the past decade.

There is good evidence that contemporary peer review process can be further improved. The American College of Obstetrics and Gynecology has offered a Voluntary Review of Quality of Care Program for more than 2 decades. Perceived issues with the adequacy of peer review were an explicit reason for requesting this service by 15% of participating hospitals, yet recommendations for improved peer review process were made to 60%. A 2007 study of peer review in US hospitals found wide variation in practice. The more effective programs had more features consistent with quality improvement principles. There were substantial opportunities for program improvement. The implication is that a new QI model for peer review seems to be evolving.

A 2009 study confirmed these findings in a separate sampling of hospitals. It also showed that important differences among programs predict a meaningful portion of the variation on 32 objective measures of patient care quality and safety.

A four-year longitudinal study of 300 programs identified the quality of case review and the likelihood of self-reporting of adverse events, near misses and hazardous conditions as additional multivariate predictors of the impact of clinical peer review on quality and safety, medical staff perceptions of the program, and clinician engagement in quality and safety initiatives. Despite a persistently high annual rate of major program change, about 80% of programs still have significant opportunity for improvement. It is argued that the out-moded QA model perpetuates a culture of blame that is toxic to efforts to advance quality and high reliability among both physicians and nurses.

External Peer Review

The 2007 study showed that the vast majority of physician peer review is done "in house": 87% of hospitals send less than 1% of their peer review cases to external agencies. The external review process is generally reserved for cases requiring special expertise for evaluation or for situations in which the independent opinion of an outside reviewer would be helpful. The process is significantly more costly than in-house review, since the majority of hospital review is done as a voluntary contribution of the medical staff.

Mandated external peer review has not played an enduring role in the US, but was tested back in the 70s. A 1972 amendment to the Social Security Act established Professional Standards Review Organizations (PSRO) with a view to controlling escalating Medicare costs through physician-organized review. The PSRO model was not considered to be effective and was replaced in 1982 by a further act of Congress which established Utilization and Quality Control Peer Review Organizations (PROs). This model too was fraught with limitations. Studies of its methods called into question its reliability and validity for peer review. A survey of Iowa state medical society members in the early 90s regarding perceptions of the PRO program illustrated the potential harm of a poorly designed program. Furthermore, the Institute of Medicine issued a report identifying the system of care as the root cause of many instances of poor quality. As a result, in the mid-90s, the PROs changed their focus and methods; and began to de-emphasize their role as agents of external peer review. The change was completed by 2002, when they were renamed Quality Improvement Organizations.

Nursing

Nursing peer review appears to have gained momentum as a result of growth of hospital participation in the American Nursing Association's Magnet Program. Even so, less than 7% of U.S. hospitals have qualified. Magnet hospitals are required to have had a peer review evaluation process in place designed to improve practice and performance for all RNs for at least 2 years. The literature on nursing peer review is more limited than that which has been developed for physician peer review, and has focused more on annual performance appraisal than on case review. No aggregate studies of clinical nursing peer review practices have been published. Nevertheless, more sophisticated studies have been reported.

Mostly what is mistakenly referred to as "peer review" in clinical practice is really a form of the annual performance evaluation. The annual performance review is a managerial process and does not meet the definition or outcomes needed related to peer review. Other organizational practices may violate the peer review guidelines set forth 1988 by the ANA 1988. The most frequent violation is the performance of direct care peer review by managers. One of the reasons for the confusion is that the ANA guidelines for peer review had been out of print prior to being reprinted and updated in 2011.

The early ANA Peer Review Guidelines (1988) and Code of Ethics for Nurses (2001) focus on maintaining standards of nursing practice and upgrading nursing care in three contemporary focus areas for peer review. The three dimensions of peer review are: (a) quality and safety, (b) role actualization, and (c) practice advancement. Each area of contemporary peer review has an organizational, unit, and individual focus. The following six peer review practice principles stem from and are grounded in the 1988 ANA Guidelines and may help to assure an evidence-based and consistent approach to peer review: 1. A peer is someone of the same rank. 2. Peer review is practice focused. 3. Feedback is timely, routine and a continuous expectation. 4. Peer review fosters a continuous learning culture of patient safety and best practice. 5. Feedback is not anonymous. 6. Feedback incorporates the developmental stage of the nurse.

Written and standardized operating procedures for peer review also need development and adoption by the direct care staff and incorporation into the professional practice model (shared governance) bylaws.

Confusion exists about the differences between the Professional Peer Review process, the Annual Performance Review (APR) and the role of peer evaluation. The APR is a managerial human resource function performed with direct reports, and is aimed at defining, aligning and recognizing each employee's contribution to the organization's success. In contrast, professional peer review is conducted within the professional practice model and is not a managerial accountability. Peer evaluation is the process of getting feedback on one's specific role competencies or "at work" behaviors from people that one works within the department and from other departments. "Colleague evaluation" is a more appropriate term than "peer evaluation" as this is not a form of professional peer review.

Composition of Peer Review Boards

There is no one standard composition of Medical peer review bodies, nor are there different names for peer review bodies of varying constituent parts. They may be carried out by State medical boards (with different standards for membership), hospital administration, senior staff, department heads, etc., or a combination of these.

State medical boards conduct peer review of licentiates, composed of physicians only or including attorneys and other non-physicians, varying by state. Physicians may be board members in primarily advisory capacities. Medical peer review may be carried out by committees that may include physicians not on the board. The same is true of state boards run by physicians from that state; board physicians or physicians unaffiliated with the board may be in medical peer review committees.

In hospitals, only a peer review committee authorized by the physician medical staff is authorized to take action regarding a physician's medical privileges at that institution. A committee convened by the hospital administration or other group within the hospital may make disciplinary recommendations to the physician medical staff.

Departmental peer review committees are composed of physicians, while hospital-based performance-appraisal and systems-analysis committees may include nurses or administrators with or without the participation of physicians.

Although medical staff bodies utilize hospital attorneys and hospital funds to try peer review cases, the California Medical Association discourages this practice; California legislation requires separation of the hospital and medical staff.

Nursing professionals have historically been less likely to participate or be subject to Peer Review. This is changing, as is the previously limited extensiveness (for example, no aggregate studies of clinical nursing peer review practices had been published as of 2010) of the literature on nursing peer review

In response to the Health Care Quality Improvement Act of 1987, (HCQIA) (P.L. 99-660) national medical associations' executives and health care organizations formed the non-profit American Medical Foundation for Peer Review and Education to provide independent assessment of medical care.

Abuse

Sham peer review is a name given to the abuse of a medical peer review process to attack a doctor for personal or other non-medical reasons. State medical boards have withheld medical records from court to frame innocent physicians as negligent. Another type of review similar to sham peer review is "incompetent peer review," in which the reviewers are unable to accurately assess the quality of care provided by their colleagues.

Controversy exists over whether medical peer review has been used as a competitive weapon in turf wars among physicians, hospitals, HMOs, and other entities and whether it is used in retaliation for whistleblowing. Many medical staff laws specify guidelines for the timeliness of peer review, in compliance with JCAHO standards, but state medical boards are not bound by such timely peer review and occasionally litigate cases for more than five years. Abuse is also referred to as "malicious peer review" by those who consider it endemic, and they allege that the creation of the National Practitioner

Data Bank under the 1986 Healthcare Quality Improvement Act (HCQIA) facilitates such abuse, creating a 'third-rail' or a 'first-strike' mentality by granting significant immunity from liability to doctors and others who participate in peer reviews.

The American Medical Association conducted an investigation of medical peer review in 2007 and concluded that while it is easy to allege misconduct, proven cases of malicious peer review are rare. Parenthetically, it is difficult to prove wrongdoing on behalf of a review committee that can use their clinical and administrative privileges to conceal exculpatory evidence.

The California legislature framed its statutes so as to allow that a peer review can be found in court to have been improper due to bad faith or malice, in which case the peer reviewers' immunities from civil liability "fall by the wayside".

Dishonesty by healthcare institutions is well-described in the literature and there is no incentive for those that lie to the public about patient care to be honest with a peer review committee.

Cases of alleged sham peer review are numerous and include cases such as Khajavi v. Feather River Anesthesiology Medical Group, Mileikowsky v. Tenet, and Roland Chalifoux.

Defenders of the Health Care Quality Improvement Act state that the National Practitioner Data Bank protects patients by helping preventing errant physicians who have lost their privileges in one state from traveling to practice in another state. Physicians who allege they have been affected by sham peer review are also less able to find work when they move to another state, as Roland Chalifoux did. Moreover, neither opponents or supporters of the NPDB can be completely satisfied, as Chalifoux' case shows that just as physicians who were unjustly accused may be deprived of work in this way, those who have erred might still find work in other states.

References

- Haralambos & Holborn. Sociology: Themes and perspectives (2004) 6th ed, Collins Educational. ISBN 978-0-00-715447-0.
- Given, Lisa M. (2008). The Sage encyclopedia of qualitative research methods. Los Angeles, Calif.: Sage Publications. ISBN 1-4129-4163-6.
- Rescuing Science from Politics: Regulation and the Distortion of Scientific ... - Google Books. Books.google.com. 2006-07-24. ISBN 9780521855204. Retrieved 2012-08-07.
- de Jager, Marije. Journal copy-editing in a non-anglophone environment. In: Matarese, Valerie (ed) (2013). Supporting Research Writing: Roles and challenges in multilingual settings. Oxford: Chandos. pp. 157–171. ISBN 1843346664.
- Rena Steinzor (July 24, 2006). "Rescuing Science from Politics". google.com. Cambridge University Press. p. 304. ISBN 0521855209.
- Martin, Brian (1997). "Suppression Stories". Fund for Intellectual Dissent. Wollongong: Fund for Intellectual Dissent. ISBN 0-646-30349-X.

- Drew, David. Little Stories of Life and Death @NHSwhistleblowr. Matador. ISBN 9781783065233. Retrieved 26 April 2016.
- Titcombe, James (November 30, 2016). Anderson-Wallace, Murray, ed. Joshua's Story: Uncovering the Morecambe Bay NHS Scandal. Anderson Wallace Publishing. p. 250. ISBN 9780993449208.
- Larivière, Vincent; Haustein, Stefanie; Mongeon, Philippe (10 June 2015). "The oligopoly of academic publishers in the digital era". PLOS ONE. 10 (6). doi:10.1371/journal.pone.0127502. Retrieved 12 May 2016.
- Zaken, Ministerie van Buitenlandse. "All European scientific articles to be freely accessible by 2020". english.eu2016.nl. Retrieved 2016-05-28.
- "AAUP Membership Benefits and Eligibility". Association of American University Presses. Retrieved August 3, 2016.
- Nosek, Brian A; Lakens, Daniel (2014). "Registered Reports: A Method to Increase the Credibility of Published Results". Social Psychology. 45 (3): 137–141. doi:10.1027/1864-9335/a000192. Retrieved 20 November 2016.
- "Mishler v. State Bd. of Med. Examiners, 849 P.2d 291, 109 Nev. 287 (1993).". 1993. p. 291. Retrieved 26 April 2016.
- "Mishler v. Nevada State Bd. of Medical Examiners, 896 F.2d 408 (9th Cir. 1990).". 1990. p. 408. Retrieved 26 April 2016.

4

Methods of Research

The methods used in research are ethnography, grounded theory, field research and meta-analysis. Ethnography is the study of people and of culture. This study has a holistic approach and also studies popular subjects such as sociology, history and communication studies. The section discusses the methods of research in a critical manner providing key analysis to the subject matter.

Ethnography

Ethnography is the systematic study of people and cultures. It is designed to explore cultural phenomena where the researcher observes society from the point of view of the subject of the study. An ethnography is a means to represent graphically and in writing the culture of a group. The word can thus be said to have a "double meaning", which partly depends on whether it is used as a count noun or uncountably. The resulting field study or a case report reflects the knowledge and the system of meanings in the lives of a cultural group.

Ethnography, as the presentation of empirical data on human societies and cultures, was pioneered in the biological, social, and cultural branches of anthropology, but it has also become popular in the social sciences in general—sociology, communication studies, history—wherever people study ethnic groups, formations, compositions, resettlements, social welfare characteristics, materiality, spirituality, and a people's ethnogenesis. The typical ethnography is a holistic study and so includes a brief history, and an analysis of the terrain, the climate, and the habitat. In all cases it should be reflexive, make a substantial contribution toward the understanding of the social life of humans, have an aesthetic impact on the reader, and express a credible reality. An ethnography records all observed behavior and describes all symbol-meaning relations, using concepts that avoid causal explanations.

History and Meaning

The word 'ethnography' is derived from the Greek *ethnos*, meaning "a company, later a people, nation" and -graphy meaning "field of study". Ethnographic studies focus on large cultural groups of people who interact over time. Ethnography is a qualitative design, where the researcher explains about shared learnt patterns of values, behaviour, beliefs, and language of a culture shared by a group of people.

The field of anthropology originated from Europe and England designed in late 19th century. It spread its roots to the United States at the beginning of the 20th century. Some of the main contributors like EB Tylor (1832-1917) from Britain and Lewis H Morgan (1818-1881), an American scientist were considered as founders of cultural and social dimensions. Franz Boas (1858-1942), Bronislaw Malinowski (1858—1942), Ruth Benedict and Margaret Mead (1901-1978), were a group of researchers from United States who contributed the idea of cultural relativism to the literature. Boas's approach focused on the use of documents and informants, whereas, Malinowski stated that a researcher should be engrossed with the work for long periods in the field and do a participant observation by living with the informant and experiencing their way of life. He gives the view point of the native and this became the origin of field work and field methods.

Since Malinowski was very firm with his approach he applied it practically and travelled to Trobriand Island which was located off the eastern coast of New Guinea. He was interested in learning the language of the islanders and stayed there for a long time doing his field work. The field of ethnography became very popular in the late 19th century, as many social scientists gained an interest in studying modern society. Again, in the latter part of the 19th century, the field of anthropology became a good support for scientific formation. Though the field was flourishing it had a lot of threat to encounter. Post colonialism, the research climate shifted towards post-modernism and feminism. Therefore, the field of anthropology moved into discipline of social science.

Origins

Gerhard Friedrich Müller developed the concept of ethnography as a separate discipline whilst participating in the Second Kamchatka Expedition (1733–43) as a professor of history and geography. Whilst involved in the expedition, he differentiated *Völker-Beschreibung* as a distinct area of study. This became known as "ethnography," following the introduction of the Greek neologism *ethnographia* by Johann Friedrich Schöpperlin and the German variant by A. F. Thilo in 1767. August Ludwig von Schlözer and Christoph Wilhelm Jacob Gatterer of the University of Göttingen introduced the term into academic discourse in an attempt to reform the contemporary understanding of world history.

Herodotus known as the Father of History had significant works on the cultures of various peoples beyond the Hellenic realm such as nations in Scythia, which earned him the title "Barbarian lover" and may have produced the first ethnographic works.

Forms of Ethnography

There are different forms of ethnography: confessional ethnography; life history; feminist ethnography etc. Two popular forms of ethnography are realist ethnography and critical ethnography. (Qualitative Inquiry and Research Design, 93)

Realist ethnography: is a traditional approach used by cultural anthropologists. Characterized by Van Maanen (1988), it reflects a particular instance taken by the researcher toward the individual being studied. It's an objective study of the situation. It's composed from a third person's perspective by getting the data from the members on the site. The ethnographer stays as omniscient correspondent of actualities out of sight. The realist reports information in a measured style ostensibly uncontaminated by individual predisposition, political objectives and judgment. The analyst will give a detailed report of everyday life of the individuals under study. The ethnographer also uses standard categories for cultural description (e.g., family life, communication network). The ethnographer produces the participant's views through closely edited quotations and has the final work on how the culture is to be interpreted and presented. (Qualitative Inquiry and Research Design, 93)

Critical ethnography: is a kind of ethnographic research in which the creators advocate for the liberation of groups which are marginalized in society. Critical researchers typically are politically minded people who look to take a stand of opposition to inequality and domination. For example, a critical ethnographer might study schools that provide privileges to certain types of students, or counselling practices that serve to overlook the needs of underrepresented groups. (Qualitative Inquiry and Research Design, 94). The important components of a critical ethnographer is to incorporate a value- laden introduction, empower people by giving them more authority, challenging the status quo, and addressing concerns about power and control. A critical ethnographer will study issues of power, empowerment, inequality inequity, dominance, repression, hegemony and victimization. (Qualitative Inquiry and Research Design, 94)

Features of Ethnographic Research

- Involves investigation of very few cases, maybe just one case, in detail.
- Often involves working with primarily unconstructed data. This data had not been coded at the point of data collection in terms of a closed set of analytic categories.
- Emphasises on exploring social phenomena rather than testing hypotheses.
- Data analysis involves interpretation of the functions and meanings of human actions. The product of this is mainly verbal explanations, where statistical analysis and quantification play a subordinate role.
- Methodological discussions focus more on questions about how to report findings in the field than on methods of data collection and interpretation.
- Ethnographies focus on describing the culture of a group in very detailed and complex manner. The ethnography can be of the entire group or a subpart of it.

- It involves engaging in extensive field work where data collection is mainly by interviews, symbols, artefacts, observations, and many other sources of data.

- The researcher in ethnography type of research, looks for patterns of the groups mental activities, that is their ideas and beliefs expressed through language or other activities, and how they behave in their groups as expressed through their actions that the researcher observed.

Procedures for Conducting Ethnography

- Determine if ethnography is the most appropriate design to use to study the research problem. Ethnography is suitable if the needs are to describe how a cultural group works and to explore their beliefs, language, behaviours and also issues faced by the group, such as power, resistance and dominance. (Qualitative Inquiry and Research Design, 94)

- Then identify and locate a culture sharing group to study. This group is one whose members have been together for an extended period of time, so that their shared language, patterns of behaviour and attitudes have merged into discernible patterns. This group can also be a group that has been marginalized by society. (Qualitative Inquiry and Research Design, 94)

- Select cultural themes, issues or theories to study about the group. These themes, issues and theories provide an orienting framework for the study of the culture-sharing group. As discussed by Hammersley and Atkinson (2007), Wolcott (1987, 1994b, 2008-1), and Fetterman (2009). The ethnographer begins the study by examining people in interaction in ordinary settings and discerns pervasive patterns such as life cycles, events and cultural themes. (Qualitative Inquiry and Research Design, 94-95)

- For studying cultural concepts, determine which type of ethnography to use. Perhaps how the group works need to be described, or a critical ethnography can expose issues such as power, hegemony and advocacy for certain groups (Qualitative Inquiry and Research Design, 95)

- Should collect information in the context or setting where the group works or lives. This is called fieldwork. Types of information typically needed in ethnography are collected by going to the research site, respecting the daily lives of individuals at the site and collecting a wide variety of materials. Field issues of respect, reciprocity, deciding who owns the data and others are central to ethnography (Qualitative Inquiry and Research Design, 95)

- From the many sources collected, the ethnographer analyzes the data for a description of the culture-sharing group, themes that emerge from the group and an overall interpretation (Wolcott, 1994b). The researcher begins to compile a detailed description of the culture-sharing group, by focusing on a single event, on several activities, or on the group over a prolonged period of time.

- Forge a working set of rules or generalisations as to how the culture sharing group works as the final product of this analysis. The final product is a holistic cultural portrait of the group that incorporates the views of the participants (emic) as well as the views of the researcher (etic). It might also advocate for the needs of the group or suggest changes in society. (Qualitative Inquiry and Research Design, 96)

Ethnography as Method

The ethnographic method is different from other ways of conducting social science approach due to the following reasons:

- It is field-based. It is conducted in the settings in which real people actually live, rather than in laboratories where the researcher controls the elements of the behaviours to be observed or measured.
- It is personalized. It is conducted by researchers who are in day-to day, face-to-face contact with the people they are studying and who are thus both participants in and observers of the lives under study.
- It is multifactorial. It is conducted through the use of two or more data collection techniques - which may be qualitative or quantitative in nature - in order to get a conclusion.
- It requires a long term commitment i.e. it is conducted by researcher who intends to interact with people they are studying for an extended period of time. The exact time frame can vary from several weeks to a year or more.
- It is inductive. It is conducted in such a way to use an accumulation of descriptive detail to build toward general patterns or explanatory theories rather than structured to test hypotheses derived from existing theories or models.
- It is dialogic. It is conducted by a researcher whose interpretations and findings may be expounded on by the study's participants while conclusions are still in the process of formulation.
- It is holistic. It is conducted so as to yield the fullest possible portrait of the group under study.

Data Collection Methods

According to the leading social scientist, John Brewer, data collection methods are meant to capture the "social meanings and ordinary activities" of people (informants) in "naturally occurring settings" that are commonly referred to as "the field." The goal is to collect data in such a way that the researcher imposes a minimal amount of personal bias on the data. Multiple methods of data collection may be employed to facilitate a relationship that allows for a more personal and in-depth portrait of the informants and their community. These can include participant observation, field notes, interviews, and surveys.

Izmir Ethnography Museum (İzmir Etnografya Müzesi), Izmir, Turkey, from the courtyard

Ethnography museum, Budapest, Hungary

Interviews are often taped and later transcribed, allowing the interview to proceed unimpaired of note-taking, but with all information available later for full analysis. Secondary research and document analysis are also used to provide insight into the research topic. In the past, kinship charts were commonly used to "discover logical patterns and social structure in non-Western societies". In the 21st century, anthropology focuses more on the study of people in urban settings and the use of kinship charts is seldom employed.

In order to make the data collection and interpretation transparent, researchers creating ethnographies often attempt to be "reflexive". Reflexivity refers to the researcher's aim "to explore the ways in which [the] researcher's involvement with a particular study influences, acts upon and informs such research". Despite these attempts of reflexivity, no researcher can be totally unbiased. This factor has provided a basis to criticize ethnography.

Traditionally, the ethnographer focuses attention on a community, selecting knowledgeable informants who know the activities of the community well. These informants are typically asked to identify other informants who represent the community, often using snowball or chain sampling. This process is often effective in revealing common cultural denominators connected to the topic being studied. Ethnography relies greatly on up-close, personal experience. Participation, rather than just observation, is one of the keys to this process. Ethnography is very useful in social research.

Ybema et al. (2010) examine the ontological and epistemological presuppositions underlying ethnography. Ethnographic research can range from a realist perspective, in which behavior is observed, to a constructivist perspective where understanding is socially constructed by the researcher and subjects. Research can range from an objectivist account of fixed, observable behaviors to an interpretivist narrative describing "the interplay of individual agency and social structure." Critical theory researchers address "issues of power within the researcher-researched relationships and the links between knowledge and power."

Another form of data collection is that of the "image." The image is the projection that an individual puts onto an object or abstract idea. An image can be contained within the physical world through a particular individual's perspective, primarily based on that individual's past experiences. One example of an image is how an individual views a novel after completing it. The physical entity that is the novel contains a specific image in the perspective of the interpreting individual and can only be expressed by the individual in the terms of "I can tell you what an image is by telling you what it feels like." The idea of an image relies on the imagination and has been seen to be utilized by children in a very spontaneous and natural manner. Effectively, the idea of the image is a primary tool for ethnographers to collect data. The image presents the perspective, experiences, and influences of an individual as a single entity and in consequence the individual will always contain this image in the group under study.

Differences Across Disciplines

The ethnographic method is used across a range of different disciplines, primarily by anthropologists but also occasionally by sociologists. Cultural studies, (European) ethnology, sociology, economics, social work, education, design, psychology, computer science, human factors and ergonomics, ethnomusicology, folklore, religious studies, geography, history, linguistics, communication studies, performance studies, advertising, nursing, urban planning, usability, political science, social movement, and criminology are other fields which have made use of ethnography.

Cultural and Social Anthropology

Cultural anthropology and social anthropology were developed around ethnographic research and their canonical texts, which are mostly ethnographies: e.g. *Argonauts of the Western Pacific* (1922) by Bronisław Malinowski, *Ethnologische Excursion in Johore* (1875) by Nicholas Miklouho-Maclay, *Coming of Age in Samoa* (1928) by Margaret Mead, *The Nuer* (1940) by E. E. Evans-Pritchard, *Naven* (1936, 1958) by Gregory Bateson, or "The Lele of the Kasai" (1963) by Mary Douglas. Cultural and social anthropologists today place a high value on doing ethnographic research. The typical ethnography is a document written about a particular people, almost always based at least in part on emic views of where the culture begins and ends. Using language or community boundaries to bound the ethnography is common. Ethnographies are also sometimes called "case studies." Ethnographers study and interpret culture, its universalities and its variations through ethnographic study based on fieldwork. An ethnography is a specific kind of written observational science which provides an account of a particular culture, society, or community. The fieldwork usually involves spending a year or more in another society, living with the local people and learning about their ways of life. Neophyte ethnographers are strongly encouraged to develop extensive familiarity with their subject prior to entering the field; otherwise, they may find themselves in difficult situations.

Ethnographers are participant observers. They take part in events they study because it helps with understanding local behavior and thought. Classic examples are Carol B. Stack's *All Our Kin*, Jean Briggs' *Never in Anger*, Richard Lee's *Kalahari Hunter-Gatherers*, Victor Turner's *Forest of Symbols*, David Maybry-Lewis' *Akew-Shavante Society*, E.E. Evans-Pritchard's *The Nuer*, and Claude Lévi-Strauss' *Tristes Tropiques*. Iterations of ethnographic representations in the classic, modernist camp include Joseph W. Bastien's "Drum and Stethoscope" (1992), Bartholomew Dean's recent (2009) contribution, *Urarina Society, Cosmology, and History in Peruvian Amazonia*.

Bronisław Malinowski among Trobriand tribe

Part of the ethnographic collection of the Međimurje County Museum in Croatia

A typical ethnography attempts to be holistic and typically follows an outline to include a brief history of the culture in question, an analysis of the physical geography or terrain inhabited by the people under study, including climate, and often including what biological anthropologists call habitat. Folk notions of botany and zoology are presented as ethnobotany and ethnozoology alongside references from the formal sciences. Material culture, technology, and means of subsistence are usually treated next, as they are typically bound up in physical geography and include descriptions of infrastructure. Kinship and social structure (including age grading, peer groups, gender, voluntary associations, clans, moieties, and so forth, if they exist) are typically included. Languages spoken, dialects, and the history of language change are another group of standard topics. Practices of childrearing, acculturation, and emic views on personality and values usually follow after sections on social structure. Rites, rituals, and other evidence of religion have long been an interest and are sometimes central to ethnographies, especially when conducted in public where visiting anthropologists can see them.

As ethnography developed, anthropologists grew more interested in less tangible aspects of culture, such as values, worldview and what Clifford Geertz termed the "ethos" of the culture. In his fieldwork, Geertz used elements of a phenomenological approach, tracing not just the doings of people, but the cultural elements themselves. For example, if within a group of people, winking was a communicative gesture, he sought to first determine what kinds of things a wink might mean (it might mean several things). Then, he sought to determine in what contexts winks were used, and whether, as one moved about a region, winks remained meaningful in the same way. In this way, cultural boundaries of communication could be explored, as opposed to using linguistic boundaries or notions about residence. Geertz, while still following something of a traditional ethnographic outline, moved outside that outline to talk about "webs" instead of "outlines" of culture.

Within cultural anthropology, there are several subgenres of ethnography. Beginning in the 1950s and early 1960s, anthropologists began writing "bio-confessional" ethnographies that intentionally exposed the nature of ethnographic research. Famous examples include *Tristes Tropiques* (1955) by Lévi-Strauss, *The High Valley* by Kenneth Read, and *The Savage and the Innocent* by David Maybury-Lewis, as well as the mildly fictionalized *Return to Laughter* by Elenore Smith Bowen (Laura Bohannan).

Later "reflexive" ethnographies refined the technique to translate cultural differences by representing their effects on the ethnographer. Famous examples include *Deep Play: Notes on a Balinese Cockfight* by Clifford Geertz, *Reflections on Fieldwork in Morocco* by Paul Rabinow, *The Headman and I* by Jean-Paul Dumont, and *Tuhami* by Vincent Crapanzano. In the 1980s, the rhetoric of ethnography was subjected to intense scrutiny within the discipline, under the general influence of literary theory and post-colonial/post-structuralist thought. "Experimental" ethnographies that reveal the ferment of the discipline include *Shamanism, Colonialism, and the Wild Man* by Michael Taussig, *Debating Muslims* by Michael F. J. Fischer and Mehdi Abedi, *A Space on the Side of the Road* by Kathleen Stewart, and *Advocacy after Bhopal* by Kim Fortun.

This critical turn in sociocultural anthropology during the mid-1980s can be traced to the influence of the now classic (and often contested) text, *Writing Culture: The Poetics and Politics of Ethnography*, (1986) edited by James Clifford and George Marcus. *Writing Culture* helped bring changes to both anthropology and ethnography often described in terms of being 'postmodern,' 'reflexive,' 'literary,' 'deconstructive,' or 'poststructural' in nature, in that the text helped to highlight the various epistemic and political predicaments that many practitioners saw as plaguing ethnographic representations and practices.

Where Geertz's and Turner's interpretive anthropology recognized subjects as creative actors who constructed their sociocultural worlds out of symbols, postmodernists attempted to draw attention to the privileged status of the ethnographers themselves. That is, the ethnographer cannot escape the personal viewpoint in creating an ethnographic account, thus making any claims of objective neutrality highly problematic, if not altogether impossible. In regards to this last point, *Writing Culture* became a focal point for looking at how ethnographers could describe different cultures and societies without denying the subjectivity of those individuals and groups being studied while simultaneously doing so without laying claim to absolute knowledge and objective authority. Along with the development of experimental forms such as 'dialogic anthropology,' 'narrative ethnography,' and 'literary ethnography', *Writing Culture* helped to encourage the development of 'collaborative ethnography.' This exploration of the relationship between writer, audience, and subject has become a central tenet of contemporary anthropological and ethnographic practice. In certain instances, active collaboration between the researcher(s) and subject(s) has helped blend the practice of collaboration in ethnographic fieldwork with the process of creating the ethnographic product resulting from the research.

Sociology

Sociology is another field which prominently features ethnographies. Urban sociology, Atlanta University (now Clark-Atlanta University), and the Chicago School in particular are associated with ethnographic research, with some well-known early examples being *The Philadelphia Negro* (1899) by W. E. B. Du Bois, *Street Corner Society* by

William Foote Whyte and *Black Metropolis* by St. Clair Drake and Horace R. Cayton, Jr.. Major influences on this development were anthropologist Lloyd Warner, on the Chicago sociology faculty, and to Robert Park's experience as a journalist. Symbolic interactionism developed from the same tradition and yielded such sociological ethnographies as *Shared Fantasy* by Gary Alan Fine, which documents the early history of fantasy role-playing games. Other important ethnographies in sociology include Pierre Bourdieu's work on Algeria and France.

Jaber F. Gubrium's series of organizational ethnographies focused on the everyday practices of illness, care, and recovery are notable. They include *Living and Dying at Murray Manor,* which describes the social worlds of a nursing home; *Describing Care: Image and Practice in Rehabilitation,* which documents the social organization of patient subjectivity in a physical rehabilitation hospital; *Caretakers: Treating Emotionally Disturbed Children,* which features the social construction of behavioral disorders in children; and *Oldtimers and Alzheimer's: The Descriptive Organization of Senility,* which describes how the Alzheimer's disease movement constructed a new subjectivity of senile dementia and how that is organized in a geriatric hospital. Another approach to ethnography in sociology comes in the form of institutional ethnography, developed by Dorothy E. Smith for studying the social relations which structure people's everyday lives.

Other notable ethnographies include Paul Willis's *Learning to Labour,* on working class youth; the work of Elijah Anderson, Mitchell Duneier, and Loïc Wacquant on black America, and Lai Olurode's *Glimpses of Madrasa From Africa*. But even though many sub-fields and theoretical perspectives within sociology use ethnographic methods, ethnography is not the *sine qua non* of the discipline, as it is in cultural anthropology.

Communication Studies

Beginning in the 1960s and 1970s, ethnographic research methods began to be widely used by communication scholars. As the purpose of ethnography is to describe and interpret the shared and learned patterns of values, behaviors, beliefs and language of a culture-sharing group, Harris, (1968), also Agar (1980) note that ethnography is both a process and an outcome of the research. Studies such as Gerry Philipsen's analysis of cultural communication strategies in a blue-collar, working-class neighborhood on the south side of Chicago, *Speaking 'Like a Man' in Teamsterville,* paved the way for the expansion of ethnographic research in the study of communication.

Scholars of communication studies use ethnographic research methods to analyze communicative behaviors and phenomena. This is often characterized in the writing as attempts to understand taken-for-granted routines by which working definitions are socially produced. Ethnography as a method is a storied, careful, and systematic examination of the reality-generating mechanisms of everyday life (Coulon, 1995). Ethnographic work in communication studies seeks to explain "how" ordinary methods/practices/performances construct the ordinary actions used by ordinary people in the accomplishments of their identities. This often gives the perception of trying to answer the "why" and "how come" questions of human communication. Often this type of research results in a case study or field study such as an analysis of speech patterns at a protest rally, or the way firemen communicate during "down time" at a fire station. Like anthropology scholars, communication scholars often immerse themselves, and participate in and/or directly observe the particular social group being studied.

Other Fields

The American anthropologist George Spindler was a pioneer in applying ethnographic methodology to the classroom.

Anthropologists such as Daniel Miller and Mary Douglas have used ethnographic data to answer academic questions about consumers and consumption. In this sense, Tony Salvador, Genevieve Bell, and Ken Anderson describe design ethnography as being "a way of understanding the particulars of daily life in such a way as to increase the success probability of a new product or service or, more appropriately, to reduce the probability of failure specifically due to a lack of understanding of the basic behaviors and frameworks of consumers." Sociologist Sam Ladner argues in her book, that understanding consumers and their desires requires a shift in "standpoint," one that only ethnography provides. The results are products and services that respond to consumers' unmet needs.

Businesses, too, have found ethnographers helpful for understanding how people use products and services. Companies make increasing use of ethnographic methods to understand consumers and consumption, or for new product development (such as video ethnography). The *Ethnographic Praxis in Industry* (EPIC) conference is evidence of this. Ethnographers' systematic and holistic approach to real-life experience is valued by product developers, who use the method to understand unstated desires or cultural practices that surround products. Where focus groups fail to inform marketers about what people really do, ethnography links what people say to what they do—avoiding the pitfalls that come from relying only on self-reported, focus-group data.

Evaluating Ethnography

Ethnographic methodology is not usually evaluated in terms of philosophical standpoint (such as positivism and emotionalism). Ethnographic studies need to be evaluated in some manner. No consensus has been developed on evaluation standards, but Richardson (2000, p. 254) provides five criteria that ethnographers might find helpful. Jaber F. Gubrium and James A. Holstein's (1997) monograph, *The New Language of Qualitative Method,* discusses forms of ethnography in terms of their "methods talk."

1. *Substantive contribution*: "Does the piece contribute to our understanding of social-life?"
2. *Aesthetic merit*: "Does this piece succeed aesthetically?"
3. *Reflexivity*: "How did the author come to write this text...Is there adequate self-awareness and self-exposure for the reader to make judgments about the point of view?"
4. *Impact*: "Does this affect me? Emotionally? Intellectually?" Does it move me?
5. *Expresses a reality*: "Does it seem 'true'—a credible account of a cultural, social, individual, or communal sense of the 'real'?"

Challenges of Ethnography

Ethnography, which is a method dedicated entirely to field work, is aimed at gaining a deeper insight of a certain people's knowledge and social culture.

Ethnography's advantages are:

- It can open up certain experiences during group research that other research methods fail to cover.
- Notions that are taken for granted can be highlighted and confronted.
- It can tap into intuitive and deep human understanding of and interpretations of (by the ethnographer) the accounts of informants (those who are being studied), which goes far beyond what quantitative research can do in terms of extracting meanings.
- Ethnography allows people outside of a culture (whether of a primitive tribe or of a corporation's employees) to learn about its members' practices, motives, understandings and values.

However, there are certain challenges or limitations for the ethnographic method:

- Deep expertise is required: Ethnographers must accumulate knowledge about the methods and domains of interest, which can take considerable training and time.
- Sensitivity: The ethnographer is an outsider and must exercise discretion and caution to avoid offending, alienating or harming those being observed.
- Access: Negotiating access to field sites and participants can be time-consuming and difficult. Secretive or guarded organizations may require different approaches in order for researchers to succeed.

- Duration and cost: Research can involve prolonged time in the field, particularly because building trust with participants is usually necessary for obtaining rich data.
- Bias: Ethnographers bring their own experience to bear in pursuing questions to ask and reviewing data, which can lead to biases in directions of inquiry and analysis.
- Descriptive approach: Ethnography relies heavily on story telling and the presentation of critical incidents, which is inevitably selective and viewed as a weakness by those used to the scientific approaches of hypothesis testing, quantification and replication.

Ethics

Gary Alan Fine argues that the nature of ethnographic inquiry demands that researchers deviate from formal and idealistic rules or ethics that have come to be widely accepted in qualitative and quantitative approaches in research. Many of these ethical assumptions are rooted in positivist and post-positivist epistemologies that have adapted over time, but are apparent and must be accounted for in all research paradigms. These ethical dilemmas are evident throughout the entire process of conducting ethnographies, including the design, implementation, and reporting of an ethnographic study. Essentially, Fine maintains that researchers are typically not as ethical as they claim or assume to be — and that "each job includes ways of doing things that would be inappropriate for others to know".

Fine is not necessarily casting blame at ethnographic researchers, but tries to show that researchers often make idealized ethical claims and standards which in are inherently based on partial truths and self-deceptions. Fine also acknowledges that many of these partial truths and self-deceptions are unavoidable. He maintains that "illusions" are essential to maintain an occupational reputation and avoid potentially more caustic consequences. He claims, "Ethnographers cannot help but lie, but in lying, we reveal truths that escape those who are not so bold". Based on these assertions, Fine establishes three conceptual clusters in which ethnographic ethical dilemmas can be situated: "Classic Virtues", "Technical Skills", and "Ethnographic Self".

Much debate surrounding the issue of ethics arose following revelations about how the ethnographer Napoleon Chagnon conducted his ethnographic fieldwork with the Yanomani people of South America.

While there is no international standard on Ethnographic Ethics, many western anthropologists look to the American Anthropological Association for guidance when conducting ethnographic work. In 2009 the Association adopted a code of ethics, stating: Anthropologists have "moral obligations as members of other groups, such as the family, religion, and community, as well as the profession". The code of ethics notes that anthropologists are part of a wider scholarly and political network, as well as human and natural environment, which needs to be reported on respectfully. The code of ethics recognizes that sometimes very close and personal relationship can sometimes develop from doing ethnographic work. The Association acknowledges that the code is limited in scope; ethnographic work can sometimes be multidisciplinary, and anthropologists need to be familiar with ethics and perspectives of other disciplines as well. The eight-page code of ethics outlines ethical considerations for those conducting Research, Teaching, Application and Dissemination of Results, which are briefly outlined below.

- Conducting Research-When conducting research Anthropologists need to be aware of the potential impacts of the research on the people and animals they study. If the seeking of new knowledge will negatively impact the people and animals they will be studying they may not undertake the study according to the code of ethics.
- Teaching-When teaching the discipline of anthropology, instructors are required to inform students of the ethical dilemmas of conducting ethnographies and field work.
- Application-When conducting an ethnography, Anthropologists must be "open with funders, colleagues, persons studied or providing information, and relevant parties affected by the work about the purpose(s), potential impacts, and source(s) of support for the work."
- Dissemination of Results-When disseminating results of an ethnography, "[a] nthropologists have an ethical obligation to consider the potential impact of both their research and the communication or dissemination of the results of their research on all directly or indirectly involved." Research results of ethnographies should not be withheld from participants in the research if that research is being observed by other people.

Classic Virtues

- "The kindly ethnographer" – Most ethnographers present themselves as being more sympathetic than they are, which aids in the research process, but is also deceptive. The identity that we present to subjects is different from who we are in other circumstances.
- "The friendly ethnographer" – Ethnographers operate under the assumption that they should not dislike anyone. When ethnographers find they intensely dislike individuals encountered in the research, they may crop them out of the findings.
- "The honest ethnographer" – If research participants know the research goals,

their responses will likely be skewed. Therefore, ethnographers often conceal what they know in order to increase the likelihood of acceptance by participants.

Technical Skills

- "The Precise Ethnographer" – Ethnographers often create the illusion that field notes are data and reflect what "really" happened. They engage in the opposite of plagiarism, giving undeserved credit through loose interpretations and paraphrasing. Researchers take near-fictions and turn them into claims of fact. The closest ethnographers can ever really get to reality is an approximate truth.

- "The Observant Ethnographer" – Readers of ethnography are often led to assume the report of a scene is complete – that little of importance was missed. In reality, an ethnographer will always miss some aspect because of lacking omniscience. Everything is open to multiple interpretations and misunderstandings. As ethnographers' skills in observation and collection of data vary by individual, what is depicted in ethnography can never be the whole picture.

- "The Unobtrusive Ethnographer" – As a "participant" in the scene, the researcher will always have an effect on the communication that occurs within the research site. The degree to which one is an "active member" affects the extent to which sympathetic understanding is possible.

Ethnographic Self

The following are commonly misconceived conceptions of ethnographers:

- "The Candid Ethnographer" – Where the researcher personally situates within the ethnography is ethically problematic. There is an illusion that everything reported was observed by the researcher.

- "The Chaste Ethnographer" – When ethnographers participate within the field, they invariably develop relationships with research subjects/participants. These relationships are sometimes not accounted for within the reporting of the ethnography, although they may influence the research findings.

- "The Fair Ethnographer" – Fine claims that objectivity is an illusion and that everything in ethnography is known from a perspective. Therefore, it is unethical for a researcher to report fairness in findings.

- "The Literary Ethnographer" – Representation is a balancing act of determining what to "show" through poetic/prosaic language and style, versus what to "tell" via straightforward, 'factual' reporting. The individual skills of an ethnographer influence what appears to be the value of the research.

According to Norman K. Denzin, ethnographers should consider the following eight principles when observing, recording, and sampling data:

1. The groups should combine symbolic meanings with patterns of interaction.
2. Observe the world from the point of view of the subject, while maintaining the distinction between everyday and scientific perceptions of reality.
3. Link the group's symbols and their meanings with the social relationships.
4. Record all behaviour.
5. Methodology should highlight phases of process, change, and stability.
6. The act should be a type of symbolic interactionism.
7. Use concepts that would avoid casual explanations.

Examples of Studies that Can use an Ethnographic Approach

- To study the behaviour of workers at a store in a mall - when the manager is present, and when he is not.
- To observe the kind of punishments children are given for not completing their homework at a particular school.
- To follow hygiene patterns of adolescents in a particular dormitory.
- To study altruistic behaviour members of a particular church display for each other.
- To examine health habits of sex workers from a particular locality.

Grounded Theory

Grounded theory (GT) is a systematic methodology in the social sciences involving the construction of theory through the analysis of data. Grounded theory is a research methodology which operates almost in a reverse fashion from social science research in the positivist tradition. Unlike positivist research, a study using grounded theory is likely to begin with a question, or even just with the collection of qualitative data. As researchers review the data collected, repeated ideas, concepts or elements become apparent, and are tagged with *codes*, which have been extracted from the data. As more data are collected, and as data are re-reviewed, codes can be grouped into concepts, and then into categories. These categories may become the basis for new theory. Thus, grounded theory is quite different from the traditional model of research, where the researcher chooses an existing theoretical framework, and only then collects data to show how the theory does or does not apply to the phenomenon under study.

Background

Grounded theory is a general methodology, a way of thinking about and conceptualizing data. It focuses on the studies of diverse populations from areas like remarriage after divorce (Cauhape, 1983) and Professional Socialization (Broadhed, 1983). The Grounded Theory method was developed by two sociologists, Barney Glaser and Anselm Strauss. Their collaboration in research on dying hospital patients led them to write *Awareness of Dying* in 1965. In this research they developed the constant comparative method, later known as Grounded Theory Method. There were three main purposes behind the publication of *The Discovery of Grounded Theory*:

1. Rationale of the theory to be grounded is that this theory helps close the gap between theory and empirical research.
2. Helped in suggesting the logic of grounded theories.
3. This book helped to legitimize careful qualitative research. This was seen to be the most important because, by the 1960s, quantitative research methods had taken an upper hand in the fields of research and qualitative was not seen as an adequate method of verification.

This theory mainly came into existence when there was a wave of criticism towards the fundamentalist and structuralist theories that were deductive and speculative in nature.

After two decades, sociologists and psychologists showed some appreciation for the Grounded theory because of its explicit and systematic conceptualization of the theory. *The Discovery of Grounded Theory* (1967) was published simultaneously in the United States and the United Kingdom, because of which the theory became well known among qualitative researchers and graduate students of those countries.

The turning point for this theory came after the publishing of two main monographs/ works which dealt with "dying in hospitals". This helped the theory to gain some significance in the fields of medical sociology, psychology and psychiatry. From its beginnings in health, the grounded theory method has come to prominence in fields as diverse as drama, management, manufacturing and education.

Philosophical Underpinnings

Grounded theory combines diverse traditions in sociology, positivism and symbolic interactionism as it is according to Ralph, Birks & Chapman (2015) "methodologically dynamic". Glaser's strong training in positivism enabled him to code the qualitative responses, however Strauss's training looked at the "active" role of people who live in it. Strauss recognized the profundity and richness of qualitative research regarding social processes and the complexity of social life, Glaser recognized the systematic analysis inherent in quantitative research through line by line examination, followed by the generation of codes, categories, and properties. According to Glaser (1992), the strategy of Grounded Theory is to take the interpretation of meaning in social interaction on board and study "the interrelationship between meaning in the perception of the subjects and their action". Therefore, through the meaning of symbols, human beings interpret their world and the actors who interact with them, while Grounded Theory translates and discovers new understandings of human beings' behaviors that are generated from the meaning of symbols. Symbolic interactionism is considered to be one of the most important theories to have influenced grounded theory, according to it understanding the world by interpreting human interaction, which occurs through the use of symbols, such as language. According to Milliken and Schreiber in Aldiabat and Navenec, the grounded theorist's task is to gain knowledge about the socially-shared meaning that forms the behaviors and the reality of the participants being studied.

Stages of Analysis

Once the data are collected, grounded theory analysis involves the following basic steps:

1. Coding text and theorizing: In grounded theory research, the search for the theory starts with the very first line of the very first interview that one codes. It involves taking a small chunk of the text where line by line is being coded. Useful concepts are being identified where key phrases are being marked. The concepts are named. Another chunk of text is then taken and the above-mentioned steps are being repeated. According to Strauss and Corbin, this process is called open coding and Charmaz called it initial coding. Basically, this process is breaking data into conceptual components. The next step involves a lot more theorizing, as in when coding is being done examples are being pulled out, examples of concepts together and think about how each concept can be related to a larger more inclusive concept. This involves the constant comparative method and it goes on throughout the grounding theory process, right up through the development of complete theories.

2. Memoing and theorizing: Memoing is when the running notes of each of the concepts that are being identified are kept. It is the intermediate step between the coding and the first draft of the completed analysis. Memos are field notes about the concepts in which one lays out their observations and insights. Memoing starts with the first concept that has been identified and continues right through the process of breaking the text and of building theories.

3. Integrating, refining and writing up theories: Once coding categories emerges, the next step is to link them together in theoretical models around a central cat-

> egory that hold everything together. The constant comparative method comes into play, along with negative case analysis which looks for cases that do not confirm the model. Basically one generates a model about how whatever one is studying works right from the first interview and see if the model holds up as one analyze more interviews.

Theorizing is involved in all these steps. One is required to build and test theory all the way through till the end of a project.

Premise

Grounded theory method is a systematic generation of theory from data that contains both inductive and deductive thinking. One goal is to formulate hypotheses based on conceptual ideas. Others may try to verify the hypotheses that are generated by constantly comparing conceptualized data on different levels of abstraction, and these comparisons contain deductive steps. Another goal of a grounded theory study is to discover the participants' main concern and how they continually try to resolve it. The questions the researcher repeatedly asks in grounded theory are "What's going on?" and "What is the main problem of the participants, and how are they trying to solve it?" These questions will be answered by the core variable and its subcores and properties in due course.

Grounded theory method does not aim for the "truth" but to conceptualize what is going on by using empirical research. In a way, grounded theory method resembles what many researchers do when retrospectively formulating new hypotheses to fit data. However, when applying the grounded theory method, the researcher does not formulate the hypotheses in advance since preconceived hypotheses result in a theory that is ungrounded from the data.

If the researcher's goal is accurate description, then another method should be chosen since grounded theory is not a descriptive method. Instead it has the goal of generating concepts that explain the way that people resolve their central concerns regardless of time and place. The use of description in a theory generated by the grounded theory method is mainly to illustrate concepts.

In most behavioral research endeavors, persons or patients are units of analysis, whereas in GT the unit of analysis is the incident. Typically several hundred incidents are analyzed in a grounded theory study since usually every participant reports many incidents.

When comparing many incidents in a certain area, the emerging concepts and their relationships are in reality probability statements. Consequently, GT is a general method that can use any kind of data even though the most common use is with qualitative data (Glaser, 2001, 2003). However, although working with probabilities, most GT studies are considered as qualitative since statistical methods are not used, and figures are not presented. The results of GT are not a reporting of statistically significant probabilities but a set of probability statements about the relationship between concepts, or an integrated set of conceptual hypotheses developed from empirical data (Glaser 1998). Validity in its traditional sense is consequently not an issue in GT, which instead should be judged by fit, relevance, workability, and modifiability (Glaser & Strauss 1967, Glaser 1978, Glaser 1998).

Fit has to do with how closely concepts fit with the incidents they are representing, and this is related to how thorough the constant comparison of incidents to concepts was done.

Relevance. A relevant study deals with the real concern of participants, evokes "grab" (captures the attention) and is not only of academic interest.

Workability. The theory works when it explains how the problem is being solved with much variation.

Modifiability. A modifiable theory can be altered when new relevant data are compared to existing data. A GT is never right or wrong, it just has more or less fit, relevance, workability and modifiability.

Nomenclature

A concept is the overall element and includes the categories which are conceptual elements standing by themselves, and properties of categories, which are conceptual aspects of categories (Glaser & Strauss, 1967). The core variable explains most of the participants' main concern with as much variation as possible. It has the most powerful properties to picture what's going on, but with as few properties as possible needed to do so. A popular type of core variable can be theoretically modeled as a basic social process that accounts for most of the variation in change over time, context, and behavior in the studied area. "GT is multivariate. It happens sequentially, subsequently, simultaneously, serendipitously, and scheduled" (Glaser, 1998).

All is data is a fundamental property of GT which means that everything that gets in the researcher's way when studying a certain area is data. Not only interviews or observations but anything are data that help the researcher generating concepts for the emerging theory. According to Ralph, Birks & Chapman (2014) field notes can come from informal interviews, lectures, seminars, expert group meetings, newspaper articles, Internet mail lists, even television shows, conversations with friends etc. Another linked with this concept technique consists of conducting self interview and treating that interview like any other data, coding and comparing it to other data and generating concepts from it.

Open coding or substantive coding is conceptualizing on the first level of abstraction. Written data from field notes or transcripts are conceptualized line by line. In the beginning of a study everything is coded in order to find out about the problem and how it is being resolved. The coding is often done in the margin of the field notes. This phase is often tedious since it involves conceptualizing all the incidents in the data, which yields many concepts. These are compared as more data is coded, merged into new concepts, and eventually renamed and modified. The GT researcher goes back and forth while comparing data, constantly modifying, and sharpening the growing theory at the same time as she follows the build-up schedule of GT's different steps.

> *On a related note,* Strauss and Corbin (1990, 1998) also proposed axial coding and defined it in 1990 as "a set of procedures whereby data are put back together in new ways after open coding, by making connections between categories." They proposed a "coding paradigm" (also discussed, among others, by Kelle, 2005) that involved "conditions, context, action/ interactional strategies and consequences." (Strauss & Corbin, 1990, p. 96)

Selective coding is done after having found the core variable or what is thought to be the core, the tentative core. The core explains the behavior of the participants in resolving their main concern. The tentative core is never wrong. It just more or less fits with the data. After the core variable is chosen, researchers selectively code data with the core guiding their coding, not bothering about concepts with little importance to the core and its subcores. Also, they now selectively sample new data with the core in mind, which is called theoretical sampling – a deductive part of GT. Selective coding delimits the study, which makes it move fast. This is indeed encouraged while doing GT (Glaser, 1998) since GT is not concerned with data accuracy as in descriptive research but is about generating concepts that are abstract of time, place and people. Selective coding could be done by going over old field notes or memos which are already coded once at an earlier stage or by coding newly gathered data.

Theoretical codes integrate the theory by weaving the fractured concepts into hypotheses that work together in a theory explaining the main concern of the participants. Theoretical coding means that the researcher applies a theoretical model to the data. It is important that this model is not forced beforehand but has emerged during the comparative process of GT. So the theoretical codes just as substantives codes should emerge from the process of constantly comparing the data in field notes and memos.

Memoing

Theoretical memoing is "the core stage of grounded theory methodology" (Glaser 1998). "Memos are the theorizing write-up of ideas about substantive codes and their theoretically coded relationships as they emerge during coding, collecting and analyzing data, and during memoing" (Glaser 1998).

Memoing is also important in the early phase of a GT study such as open coding. The researcher is then conceptualizing incidents, and memoing helps this process. Theoretical memos can be anything written or drawn in the constant comparison that makes up a GT. Memos are important tools to both refine and keep track of ideas that develop when researchers compare incidents to incidents and then concepts to concepts in the evolving theory. In memos, they develop ideas about naming concepts and relating them to each other and try the relationships between concepts in two-by-two tables, in diagrams or figures or whatever makes the ideas flow, and generates comparative power.

Without memoing, the theory is superficial and the concepts generated are not very original. Memoing works as an accumulation of written ideas into a bank of ideas about concepts and how they relate to each other. This bank contains rich parts of what will later be the written theory. Memoing is total creative freedom without rules of writing, grammar or style (Glaser 1998). The writing must be an instrument for outflow of ideas, and nothing else. When people write memos, the ideas become more realistic, being converted from thoughts into words, and thus ideas communicable to the afterworld.

In GT the preconscious processing that occurs when coding and comparing is recognized. The researcher is encouraged to register ideas about the ongoing study that eventually pop up in everyday situations, and awareness of the serendipity of the method is also necessary to achieve good results.

Sorting

In the next step memos are sorted, which is the key to formulate the theory for presentation to others. Sorting puts fractured data back together. During sorting lots of new ideas emerge, which in turn are recorded in new memos giving the memo-on-memos phenomenon. Sorting memos generates theory that explains the main action in the studied area. A theory written from unsorted memos may be rich in ideas but the connection between concepts is weak.

Writing

Writing up the sorted memo piles follows after sorting, and at this stage the theory is close to the written GT product. The different categories are now related to each other and the core variable. The theoretical density should be dosed so that concepts are mixed with description in words, tables, or figures to optimize readability.

In the later rewriting the relevant literature is woven in to put the theory in a scholarly context. Finally, the GT is edited for style and language and eventually submitted for publication.

No Pre-research Literature Review, No Taping and No Talk

GT according to Glaser gives the researcher freedom to generate new concepts explaining human behavior. This freedom is optimal when the researcher refrains from taping interviews, doing a pre-research literature review, and talking about the research before it is written up. These rules makes GT different from most other methods using qualitative data.

No pre-research literature review. Studying the literature of the area under study gives preconceptions about what to find and the researcher gets desensitized by borrowed concepts. Instead, the GT method increases theoretical sensitivity. The literature should instead be read in the sorting stage being treated as more data to code and compare with what has already been coded and generated.

No taping. Taping and transcribing interviews is common in qualitative research, but is counter-productive and a waste of time in GT which moves fast when the researcher delimits her data by field-noting interviews and soon after generates concepts that fit with data, are relevant and work in explaining what participants are doing to resolve their main concern. However, Kathy Charmaz counters this point, insisting that transcribing, coding, and re-coding are integral to the development of the theory.

No talk. Talking about the theory before it is written up drains the researcher of motivational energy. Talking can either render praise or criticism, and both diminish the motivational drive to write memos that develop and refine the concepts and the theory (Glaser 1998). Positive feedback makes researchers content with what they have and negative feedback hampers their self-confidence. Talking about the GT should be restricted to persons capable of helping the researcher without influencing her final judgments.

Split in Methodology and Methods

Ralph, Birks & Chapman (2015) explain the split in divergence grounded theory methodology in the article "The Methodological Dynamism of Grounded Theory" and how grounded theory has been influenced by varying schools of thought over the years.

Divergence

Since their original publication in 1967, Glaser and Strauss have disagreed on how to apply the grounded theory method, resulting in a split between Straussian and Glaserian paradigms. This split occurred most obviously after Strauss published *Qualitative Analysis for Social Scientists* (1987). Thereafter Strauss, together with Juliet Corbin, published *Basics of Qualitative Research: Grounded Theory Procedures and Techniques* in 1990. This was followed by a rebuke by Glaser (1992) who set out, chapter by chapter, to highlight the differences in what he argued was original grounded theory and why, according to Glaser, what Strauss and Corbin had written was not grounded theory in its "intended form" but was rather a form of qualitative data analysis. This divergence in methodology is a subject of much academic debate, which Glaser (1998) calls a "rhetorical wrestle". Glaser continues to write about and teach the original grounded theory method.

According to Kelle (2005), "the controversy between Glaser and Strauss boils down to the question of whether the researcher uses a well-defined 'coding paradigm' and always looks systematically for 'causal conditions,' 'phenomena/context, intervening conditions, action strategies' and 'consequences' in the data, or whether theoretical codes are employed as they emerge in the same way as substantive codes emerge, but drawing on a huge fund of 'coding families.' Both strategies have their pros and cons. Novices who wish to get clear advice on how to structure data material may be satisfied with the use of the coding paradigm. Since the paradigm consists of theoretical terms which carry only limited empirical content the risk is not very high that data are forced by its application. However, it must not be forgotten that it is linked to a certain micro-sociological perspective. Many researchers may concur with that approach especially since qualitative research always had a relation to micro-sociological action theory, but others who want to employ a macro-sociological and system theory perspective may feel that the use of the coding paradigm would lead them astray."

Glaser's Approach

Glaser originated the basic process of Grounded theory method described as the constant comparative method where the analyst begins analysis with the first data collected and constantly compares indicators, concepts and categories as the theory emerges.

The first book, *The Discovery of Grounded Theory*, published in 1967, was "developed in close and equal collaboration" by Glaser and Strauss. Glaser wrote "Theoretical Sensitivity" in 1978 and has since written five more books on the method and edited five readers with a collection of grounded theory articles and dissertations.

The Glaserian method is *not* a qualitative research method, but claims the dictum "all is data". This means that not only interview or observational data but also surveys or statistical analyses or "whatever comes the researcher's way while studying a substantive area" (Glaser quote) can be used in the comparative process as well as literature data from science or media or even fiction. Thus the method according to Glaser is not limited to the realm of qualitative research, which he calls "QDA" (Qualitative Data Analysis). QDA is devoted to descriptive accuracy while the Glaserian method emphasizes conceptualization abstract of time, place and people. A theory discovered with the grounded theory method should be easy to use outside of the substantive area where it was generated.

Strauss and Corbin's Approach

Generally speaking, grounded theory is an approach for looking systematically at (mostly) qualitative data (like transcripts of interviews or protocols of observations) aiming at the generation of theory. Sometimes, grounded theory is seen as a qualitative method, but grounded theory reaches farther: it combines a specific style of research (or a paradigm) with pragmatic theory of action and with some methodological guidelines.

This approach was written down and systematized in the 1960s by Anselm Strauss (himself a student of Herbert Blumer) and Barney Glaser (a student of Paul Lazarsfeld), while working together in studying the sociology of illness at the University of California, San Francisco. For and with their studies, they developed a methodology, which was then made explicit and became the foundation stone for an important branch of qualitative sociology.

Important concepts of grounded theory method are categories, codes and codings. The research principle behind grounded theory method is neither inductive nor deductive, but combines both in a way of abductive reasoning (coming from the works of Charles Sanders Peirce). This leads to a research practice where data sampling, data analysis and theory development are not seen as distinct and disjunct, but as different steps to be repeated until one can describe and explain the phenomenon that is to be researched. This stopping point is reached when new data does not change the emerging theory anymore.

In an interview that was conducted shortly before Strauss' death (1994), he named three basic elements every grounded theory approach should include (Legewie/Schervier-Legewie (2004)). These three elements are:

- *Theoretical sensitive coding*, that is, generating theoretical strong concepts from the data to explain the phenomenon researched;
- *theoretical sampling*, that is, deciding whom to interview or what to observe next according to the state of theory generation, and that implies starting data analysis with the first interview, and writing down memos and hypotheses early;
- the need to *compare* between phenomena and contexts to make the theory strong.

Differences

Grounded theory method according to Glaser emphasizes induction or emergence, and the individual researcher's creativity within a clear frame of stages, while Strauss is more interested in validation criteria and a systematic approach.

Constructivist

A later version of GT called constructivist GT, which was rooted in pragmatism and relativist epistemology, assumes that neither data nor theories are discovered, but are constructed by the researcher as a result of his or her interactions with the field and its participants. Data are co-constructed by researcher and participants, and colored by the researcher's perspectives, values, privileges, positions, interactions, and geographical locations. This position takes a middle ground between the realist and postmodernist positions by assuming an "obdurate reality" at the same time as it assumes multiple realities and multiple perspectives on these realities. Within this approach, a literature review is used in a constructive and data-sensitive way without forcing it on data.

Use in Various Disciplines

Grounded theory is "shaped by the desire to discover social and psychological processes" However grounded theory is not restricted to these two disciplines of study. As Gibbs points out, the process of grounded theory can be and has been applied to a number of different disciplines such as medicine, law, and economics to name a few. Grounded theory has gone global among the disciplines of nursing, business, and education and less so among other social-psychological-oriented disciplines such as social welfare, psychology, sociology, and art.

Grounded theory focuses more on the procedure and not on the discipline. Rather than being limited to a particular discipline or form of data collection, grounded theory has been found useful across multiple research areas (Wells 1995). Here are some examples:

1. In psychology, grounded theory is used to understand the role of therapeutic distance for adult clients with attachment anxiety.
2. In sociology, grounded theory is used to discover the meaning of spirituality in cancer patients, and how their beliefs influence their attitude towards cancer treatments.
3. Public health researchers have used grounded theory to examine nursing home preparedness needs through the experiences of Hurricane Katrina refugees sheltered in nursing homes.
4. In business, grounded theory is used by managers to explain the ways in which organizational characteristics explain co-worker support.
5. In software engineering, grounded theory has been used to study daily stand-up meetings.
6. Grounded theory has also helped research in the field of Information Technology to study the use of computer technology in older adults.
7. In nursing, grounded theory has been used to examine how bedside shift report can be used to keep patients safe

Benefits

The benefits of using grounded theory include:

Ecological validity: Ecological validity is the extent to which research findings accurately represent real-world settings. Grounded theories are usually ecologically valid because they are similar to the data from which they were established. Although the constructs in a grounded theory are appropriately abstract (since their goal is to explain other similar phenomenon), they are context-specific, detailed, and tightly connected to the data.

Novelty: Because grounded theories are not tied to any preexisting theory, grounded theories are often fresh and new and have the potential for innovative discoveries in science and other areas.

Parsimony: Parsimony involves using the simplest possible definition to explain complex phenomenon. Grounded theories aim to provide practical and simple explanations about complex phenomena by converting them into abstract constructs and hypothesizing their relationships. They offer helpful and relatively easy-to-remember layouts for us to understand our world a little bit better.

Grounded theory has further significance because:

1. It provides explicit, sequential guidelines for conducting qualitative research.
2. It offers specific strategies for handling the analytic phases of inquiry.
3. It streamlines and integrates data collection and analysis and
4. It legitimizes qualitative research as scientific inquiry.

Grounded theory methods have earned their place as a standard social research method and have influenced researchers from varied disciplines and professions.

Criticisms

Critiques of grounded theory have focused on:

1. Its misunderstood status as theory (is what is produced really 'theory'?),
2. The notion of 'ground' (why is an idea of 'grounding' one's findings important in qualitative inquiry—what are they 'grounded' *in*?)
3. The claim to use and develop inductive knowledge.

These criticisms are summed up by Thomas and James. These authors also suggest that it is impossible to free oneself of preconceptions in the collection and analysis of data in the way that Glaser and Strauss say is necessary. They also point to the formulaic nature of grounded theory method and the lack of congruence of this with open and creative interpretation – which ought to be the hallmark of qualitative inquiry. They suggest that the one element of grounded theory worth keeping is constant comparative method.

Goldthorpe has put forth some criticisms of grounded theory as an effort to synthesize variables oriented as empirical studies and radical choice theory. Grounded theory allows for modifications in the formulated hypotheses at the start of the empirical research process. In grounded theory, researchers engage in excessive conceptualization and defend this as "sensitivity to context." Because of this, convergent conceptualization becomes impossible.[?] As a result of these two arguments, grounded theory escapes the testing of theory. There is a very thin line between context and regularities. Goldthorpe supports this criticism in a review of three overlapping literatures: historical sociology, comparative macrosociology, and ethnography. On the one hand, historical sociology is good at analyzing long term processes of structural change, but on the other hand, its reliance on secondary sources opens several possibilities of bias. Comparative macro-sociology may be able to contextualize with reference to institutions and historical path-dependencies, but its focus on constellations of singular causal forces makes it difficult to break with long outdated mechanical models of reasoning. Ethnography may closely analyse actual mechanisms of interaction, but it doesn't provide acceptable knowledge about underlying generative processes, since it is unable to deal with variation within and across locales. Goldthorpe's core arguments are in terms of rational action theory and probabilistic statistical models. Grounded Theory can be reductive in the search for general patterns across a population, and even the selective coding process does not fully cover the contextual issues.

The grounded theory approach can be criticized as being empiricist; that it relies too heavily on the empirical data. Considers the fieldwork data as the source of its theories and sets itself against the use of the preceding theories Strauss's version of grounded theory has been criticized in several ways-

- Grounded theory focuses on a quasi objective centered researcher with an emphasis on hypotheses, variables, reliability and replicability. This is contradictory with the more away from this more quantitative form of terminology in recent qualitative research approaches.
- It will not be appropriate to ignore the existing theories by paying less attention to the review of literature. The researcher invariably comes to the research topic by finding more about his or her own discipline.
- Grounded theory focuses more on complex methods and confusing, overlapping terminologies rather than the data. Few processes like 3 stage process with associated data fragmentation may lead the researcher to lose the track of the overall picture which is emerging.
- Poorly put forth theoretical explanations tends to be the outcome where data are linked conceptually and early to existing frameworks. Concept generation rather than the formal theory may be the best outcome. (Grbich, 2007)

Grounded theory method was developed in a period when other qualitative methods were often considered unscientific. It achieved wide acceptance of its academic rigor. Thus, especially in American academia, qualitative research is often equated to grounded theory method. This equation is sometimes criticized by qualitative researchers using other methodologies (for example, traditional ethnography, narratology, and storytelling).

One alternative to grounded theory is engaged theory. It puts an equal emphasis on doing on-the-ground work linked to analytical processes of empirical generalization. However, unlike grounded theory, engaged theory is in the critical theory tradition, locating those processes within a larger theoretical framework that specifies different levels of abstraction at which one can make claims about the world.

Field Research

Biologists collecting information in the field

Field research or fieldwork is the collection of information outside a laboratory, library or workplace setting. The approaches and methods used in field research vary across disciplines. For example, biologists who conduct field research may simply observe animals interacting with their environments, whereas social scientists conducting field research may interview or observe people in their natural environments to learn their languages, folklore, and social structures.

Field research involves a range of well-defined, although variable, methods: informal interviews, direct observation, participation in the life of the group, collective discussions, analyses of personal documents produced within the group, self-analysis, results from activities undertaken off- or on-line, and life-histories. Although the method generally is characterized as qualitative research, it may (and often does) include quantitative dimensions.

History

Field research has a long history. Cultural anthropologists have long used field research to study other cultures. Although the cultures do not have to be different, this has often been the case in the past with the study of so-called primitive cultures, and even in sociology the cultural differences have been ones of class. The work is done... in "'Fields' that is, circumscribed areas of study which have been the subject of social research". Fields could be education, industrial settings, or Amazonian rain forests. Field research may be conducted by zoologists such as Jane Goodall. Radcliff-Brown [1910] and Malinowski [1922] were early cultural anthropologists who set the models for future work.

Business use of Field research is an applied form of anthropology and is as likely to be advised by sociologists or statisticians in the case of surveys.

Consumer marketing field research is the primary marketing technique used by businesses to research their target market.

Conducting Field Research

The quality of results obtained from field research depends on the data gathered in the field. The data in turn, depend upon the field worker, his or her level of involvement, and ability to see and visualize things that other individuals visiting the area of study may fail to notice. The more open researchers are to new ideas, concepts, and things which they may not have seen in their own culture, the better will be the absorption of those ideas. Better grasping of such material means better understanding of the forces of culture operating in the area and the ways they modify the lives of the people under study. Social scientists (i.e. anthropologists, social psychologists, etc.) have always been taught to be free from ethnocentrism (i.e. the belief in the superiority of one's own ethnic group), when conducting any type of field research.

When humans themselves are the subject of study, protocols must be devised to reduce the risk of observer bias and the acquisition of too theoretical or idealized explanations of the workings of a culture. Participant observation, data collection, and survey research are examples of field research methods, in contrast to what is often called experimental or lab research.

Field Notes

When conducting field research, keeping an ethnographic record is essential to the process. Field notes are a key part of the ethnographic record. The process of field notes begin as the researcher participates in local scenes and experiences in order to make observations that will later be written up. The field researcher tries first to take mental notes of certain details in order that they be written down later.

Kinds of Field Notes

Field Note Chart

Types of Field Notes	Brief Description
Jot Notes	Key words or phrases are written down while in the field.
Field Notes Proper	A description of the physical context and the people involved, including their behavior and nonverbal communication.
Methodological Notes	New ideas that the researcher has on how to carry out the research project.
Journals and Diaries	These notes record the ethnographer's personal reactions, frustrations, and assessments of life and work in the field.

Jot Notes

The first writing that is done typically consists of jotted or condensed notes. Thus, key words or phrases are written down while the researcher is in or very close to the field. Some researchers jot field notes openly in the presence of those being studied. Adopting this practice early on enables some researchers to find that they can establish a 'note-taker' role that will be accepted or at least tolerated by those being studied. However, some researchers find that people develop expectations of what should be recorded and what should not, which can intrude upon the work being done. Other ethnographers try to avoid taking notes in the middle of scenes and experiences and instead try to place themselves on the margins of scenes and events. Others strictly avoid writing anything in the presence of those being studied. They feel that such writing can overtly remind the participants that the researcher has different commitments and priorities. Such writing can also distract the researcher from what is happening in the immediate scene in which he or she is participating. Thus, many researchers choose to make jotted notes outside the presence of those being studied. Some therefore retreat to bathrooms or stairwells in order to record field notes.

Field Notes Proper

There are three main points regarding field notes proper. First, converting jot notes into field notes should take place as soon as possible after the events take place. Secondly, field notes should be very detailed. Thus, included in field notes should be a description of the physical context and the people involved, including their behavior and nonverbal communication. Field notes should also use words that are as close as possible to the words used by the participants. Thirdly, field notes should include thoughts, impressions and explanations on the part of the researcher. In assessing the quality of field notes, the accuracy of the description and the level of detail are of utmost importance.

Methodological Notes

These notes can contain new ideas that the researcher has on how to carry out the research project. Also included can be which methods are chosen, on what basis they were chosen, how they were carried out and the outcome of such methods. Methodological notes can be kept with field notes or they can filed separately. These also serve the researcher when later writing up the methods section of a report or paper.

Journals and Diaries

Journals and diaries are written notes that record the ethnographer's personal reactions, frustrations, and assessments of life and work in the field. When constructed chronologically these journals provide a guide to the information in field notes and records. One of the most well known diaries is that of Bronislaw Malinowski regarding his research among the Trobriand Islanders. During her Pacific fieldwork Margaret Mead kept a diary and also wrote long letters to people at home which contained self-reflection that might be included in a diary.

Interviewing

Another method of data collection is interviewing, specifically interviewing in the qualitative paradigm. Interviewing can be done in different formats, this all depends on individual researcher preferences, research purpose, and the research question asked.

Analyzing Data

In qualitative research, there are many ways of analyzing data gathered in the field. One of the two most common methods of data analysis are thematic analysis and narrative analysis. As mentioned before, the type of analysis a researcher decides to use depends on the research question asked, the researcher's field, and the researcher's personal method of choice.

Field Research Across Different Disciplines

Anthropology

In anthropology, field research is organized so as to produce a kind of writing called ethnography. Ethnography can refer to both a methodology and a product of research, namely a monograph or book. Ethnography is a grounded, inductive method that heavily relies on participant-observation. Participant observation is a structured type of research strategy. It is a widely used methodology in many disciplines, particularly, cultural anthropology, but also sociology, communication studies, and social psychology. Its aim is to gain a close and intimate familiarity with a given group of individuals (such as a religious, occupational, or sub cultural group, or a particular community) and their practices through an intensive involvement with people in their natural environment, usually over an extended period of time. The method originated in field work of social anthropologists, especially the students of Franz Boas in the United States, and in the urban research of the Chicago School of sociology.

Traditional participant observation is usually undertaken over an extended period of time, ranging from several months to many years, and even generations. An extended research time period means that the researcher is able to obtain more detailed and accurate information about the individuals, community, and/or population under study. Observable details (like daily time allotment) and more hidden details (like taboo behavior) are more easily observed and interpreted over a longer period of time. A strength of observation and interaction over extended periods of time is that researchers can discover discrepancies between what participants say—and often believe—should happen (the formal system) and what actually does happen, or between different aspects of the formal system; in contrast, a one-time survey of people's answers to a set of questions might be quite consistent, but is less likely to show conflicts between different aspects of the social system or between conscious representations and behavior".

Archaeology

Field research lies at the heart of archaeological research. It may include the undertaking of broad area surveys (including aerial surveys); of more localised site surveys (including photographic, drawn, and geophysical surveys, and exercises such as field-walking); and of excavation.

Biology

In biology, field research typically involves studying of free-living wild animals in which the subjects are observed in their natural habitat, without changing, harming, or materially altering the setting or behavior of the animals under study. Field research is an indispensable part of biological science.

Animal migration tracking (including bird ringing/banding) is a frequently-used field technique, allowing field scientists to track migration patterns and routes, and animal longevity in the wild. Knowledge about animal migrations is essential to accurately determining the size and location of protected areas.

Earth and Atmospheric Sciences

In the Earth and atmospheric sciences, field research refers to field experiments (such as the VORTEX projects) utilizing in situ instruments. Permanent observation networks are also maintained for other uses but are not necessarily considered field research, nor are permanent remote sensing installations.

Economics

The objective of field research in economics is to get beneath the surface, to contrast observed behaviour with the prevailing understanding of a process, and to relate language and description to behavior (e.g. Deirdre McCloskey, 1985).

The 2009 Nobel Prize Winners in Economics, namely, Elinor Ostrom and Oliver Williamson, have advocated mixed methods and complex approaches in economics and hinted implicitly to the relevance of field research approaches in economics. In a recent interview Oliver Williamson and Elinor Ostrom discuss the importance of examining institutional contexts when performing economic analyses. Both Ostrom and Williamson agree that "top-down" panaceas or "cookie cutter" approaches to policy problems don't work. They believe that policymakers need to give local people a chance to shape the systems used to allocate resources and resolve disputes. Sometimes, Ostrom points out, local solutions can be the most efficient and effective options. This is a point of view that fits very well with anthropological research, which has for some time shown us the logic of local systems of knowledge — and the damage that can be done when "solutions" to problems are imposed from outside or above without adequate consultation. Elinor Ostrom, for example, combines field case studies and experimental lab work in her research. Using this combination, she contested longstanding assumptions about the possibility that groups of people could cooperate to solve common pool problems (as opposed to being regulated by the state or governed by the market.

Recently Swann (2008, pp. 3–5) argued that "The only way we can know something is by hearing what can be said about it by persons of every variety of opinion, and studying all modes in which it can be looked at by every character of mind'. If economist had followed Mill's wise advice, we would by now be making use of an extraordinary repertoire of research methods in applied economics, including the vernacular methods described in this book".

Edward J. Nell (1998) argued that there are two types of field research in economics. One kind can give us a carefully drawn picture of institutions and practices, general in that it applies to all activities of a certain kind of particular society or social setting, but still specialized to that society or setting. Although institutions and practices are intangibles, such a picture will be objective, a matter of fact, independent of the state of mind of the particular agents reported on. Approaching the economy from a different angle, another kind of fieldwork can give us a picture of the state of mind of economic agents (their true motivations, their beliefs, state knowledge, expectations, their preferences and values).

Public Health

In public health the use of the term field research refers to epidemiology or the study of epidemics through the gathering of data about the epidemic (such as the pathogen and vector(s) as well as social or sexual contacts, depending upon the situation).

Management

Mintzberg played a crucial role in the popularization of field research in management. The tremendous amount of work that Mintzberg put into the findings earned him the title of leader of a new school of management, the descriptive school, as opposed to the prescriptive and normative schools that preceded his work. The schools of thought derive from Taylor, Henri Fayol, Lyndall Urwick, Herbert A. Simon, and others endeavored to prescribe and expound norms to show what managers must or should do. With the arrival of Mintzberg, the question was no longer what must or should be done, but what a manager actually does during the day. More recently, in his 2004 book Managers Not MBAs, Mintzberg examined what he believes to be wrong with management education today.

Aktouf (2006, p. 198) summed-up Mintzberg observations about what takes place in the field:"First, the manager's job is not ordered, continuous, and sequential, nor is it uniform or homogeneous. On the contrary, it is fragmented, irregular, choppy, extremely changeable and variable. This work is also marked by brevity: no sooner has a manager finished one activity than he or she is called up to jump to another, and this pattern continues nonstop. Second, the manager's daily work is a not a series of self-initiated, willful actions transformed into decisions, after examining the circumstances. Rather, it is an unbroken series of reactions to all sorts of request that come from all around the manager, from both the internal and external environments. Third, the manager deals with the same issues several times, for short periods of time; he or she is far from the traditional image of the individual who deals with one problem at a time, in a calm and orderly fashion. Fourth, the manager acts as a focal point, an interface, or an intersection between several series of actors in the organization: external and internal environments, collaborators, partners, superiors, subordinates, colleagues, and so forth. He or she must constantly ensure, achieve, or facilitate interactions between all these categories of actors to allow the firm to function smoothly."

Sociology

Pierre Bourdieu played a crucial role in the popularization of fieldwork in sociology. During the Algerian War in 1958-1962, Bourdieu undertook ethnographic research into the clash through a study of the Kabyle peoples, of the Berbers laying the groundwork for his anthropological reputation. The result was his first book, Sociologie de L'Algerie (The Algerians), which was an immediate success in France and published in America in 1962. The book ("Algeria 1960: The Disenchantment of the World: The Sense of Honour: The Kabyle House or the World Reversed: Essays"), published in English in 1979 by Cambridge University Press, established him as a major figure in the field of ethnology and a pioneer advocate scholar for more intensive fieldwork in social sciences. The book was based on his decade of work as a participant-observer with the Algerian society. One of the outstanding qualities of his work has been his innovative combination of different methods and research strategies as well as his analytical skills in interpreting the obtained data.

Throughout his career, Bourdieu sought to connect his theoretical ideas with empirical research, grounded in everyday life. His work can be seen as sociology of culture. Bourdieu labeled it a "Theory of Practice". His contributions to sociology were both empirical and theoretical. His conceptual apparatus is based on three key terms, namely, habitus, capital and field. Furthermore, Bourdieu fiercely opposed Rational Choice Theory as grounded in a misunderstanding of how social agents operate. Bourdieu argued that social agents do not continuously calculate according to explicit rational and economic criteria. According to Bourdieu, social agents operate according to an implicit practical logic—a practical sense—and bodily dispositions. Social agents act according to their "feel for the game" (the "feel" being, roughly, habitus, and the "game" being the field).

Bourdieu's anthropological work was focused on the analysis of the mechanisms of reproduction of social hierarchies. Bourdieu criticized the primacy given to the economic factors, and stressed that the capacity of social actors to actively impose and engage their cultural productions and symbolic systems plays an essential role in the reproduction of social structures of domination. Bourdieu's empirical work played a crucial role in the popularization of correspondence analysis and particularly "Multiple Correspondence Analysis." Bourdieu held that these geometric techniques of data analysis are, like his sociology, inherently relational. In the preface to his book "The Craft of Sociology" Bourdieu argued that: "I use Correspondence Analysis very much, because I think that it is essentially a relational procedure whose philosophy fully expresses what in my view constitutes social reality. It is a procedure that 'thinks' in relations, as I try to do it with the concept of field."

One of the classic ethnographies in Sociology is the book Ain't No Makin' It: Aspirations & Attainment in a Low-Income Neighborhood by Jay MacLeod. The study addresses the reproduction of social inequality among low-income, male teenagers. The researcher spent time studying two groups of teenagers in a housing project in a Northeastern city of the United States. The study concludes that three different levels of analysis play their part in the reproduction of social inequality: the individual, the cultural, and the structural.

Meta-analysis

A meta-analysis is a statistical analysis that combines the results of multiple scientific studies.

The basic tenet behind meta-analyses is that there is a common truth behind all conceptually similar scientific studies, but which has been measured with a certain error within individual studies. The aim then is to use approaches from statistics to derive a pooled estimate closest to the unknown common truth based on how this error is perceived. In essence, all existing methods yield a weighted average from the results of the individual studies and what differs is the manner in which these weights are allocated and also the manner in which the uncertainty is computed around the point estimate thus generated. In addition to providing an estimate of the unknown common truth, meta-analysis has the capacity to contrast results from different studies and identify patterns among study results, sources of disagreement among those results, or other interesting relationships that may come to light in the context of multiple studies. Meta-analysis can be thought of as "conducting research about previous research." Meta-analysis can only proceed if we are able to identify a common statistical measure that is shared among studies, called the effect size, which has a standard error so that we can proceed with computing a weighted average of that common measure. Such weighting usually takes into consideration the sample sizes of the individual studies, although it can also include other factors, such as study quality.

A key benefit of this approach is the aggregation of information leading to a higher statistical power and more robust point estimate than is possible from the measure derived from any individual study. However, in performing a meta-analysis, an investigator must make choices which can affect the results, including deciding how to search for studies, selecting studies based on a set of objective criteria, dealing with incomplete data, analyzing the data, and accounting for or choosing not to account for publication bias.

Meta-analyses are often, but not always, important components of a systematic review procedure. For instance, a meta-analysis may be conducted on several clinical trials of a medical treatment, in an effort to obtain a better understanding of how well the treatment works. Here it is convenient to follow the terminology used by the Cochrane Collaboration, and use "meta-analysis" to refer to statistical methods of combining evidence, leaving other aspects of 'research synthesis' or 'evidence synthesis', such as combining information from qualitative studies, for the more general context of systematic reviews.

History

The historical roots of meta-analysis can be traced back to 17th century studies of astronomy, while a paper published in 1904 by the statistician Karl Pearson in the *British Medical Journal* which collated data from several studies of typhoid inoculation is seen as the first time a meta-analytic approach was used to aggregate the outcomes of multiple clinical studies. The first meta-analysis of all conceptually identical experiments concerning a particular research issue, and conducted by independent researchers, has been identified as the 1940 book-length publication *Extrasensory Perception After Sixty Years*, authored by Duke University psychologists J. G. Pratt, J. B. Rhine, and associates. This encompassed a review of 145 reports on ESP experiments published from 1882 to 1939, and included an estimate of the influence of unpublished papers on the overall effect (the *file-drawer problem*). Although meta-analysis is widely used in epidemiology and evidence-based medicine today, a meta-analysis of a medical treatment was not published until 1955. In the 1970s, more sophisticated analytical techniques were introduced in educational research, starting with the work of Gene V. Glass, Frank L. Schmidt and John E. Hunter.

The term "meta-analysis" was coined by Gene V. Glass, who was the first modern statistician to formalize the use of the term meta-analysis. He states *"my major interest currently is in what we have come to call ...the meta-analysis of research. The term is a bit grand, but it is precise and apt ... Meta-analysis refers to the analysis of analyses"*. Although this led to him being widely recognized as the modern founder of the method, the methodology behind what he termed "meta-analysis" predates his work by several decades. The statistical theory surrounding meta-analysis was greatly advanced by the work of Nambury S. Raju, Larry V. Hedges, Harris Cooper, Ingram Olkin, John E. Hunter, Jacob Cohen, Thomas C. Chalmers, Robert Rosenthal, Frank L. Schmidt, and Douglas G. Bonett.

Advantages

Conceptually, a meta-analysis uses a statistical approach to combine the results from multiple studies in an effort to increase power (over individual studies), improve estimates of the size of the effect and/or to resolve uncertainty when reports disagree. A meta-analysis is a statistical overview of the results from one or more systematic review. Basically, it produces a weighted average of the included study results and this approach has several advantages:

- Results can be generalized to a larger population,
- The precision and accuracy of estimates can be improved as more data is used. This, in turn, may increase the statistical power to detect an effect.
- Inconsistency of results across studies can be quantified and analyzed. For instance, does inconsistency arise from sampling error, or are study results (partially) influenced by between-study heterogeneity.
- Hypothesis testing can be applied on summary estimates,
- Moderators can be included to explain variation between studies,
- The presence of publication bias can be investigated

Problems

A meta-analysis of several small studies does not predict the results of a single large study. Some have argued that a weakness of the method is that sources of bias are not controlled by the method: a good meta-analysis of badly designed studies will still result in bad statistics. This would mean that only methodologically sound studies should be included in a meta-analysis, a practice called 'best evidence synthesis'. Other meta-analysts would include weaker studies, and add a study-level predictor variable that reflects the methodological quality of the studies to examine the effect of study quality on the effect size. However, others have argued that a better approach is to preserve information about the variance in the study sample, casting as wide a net as possible, and that methodological selection criteria introduce unwanted subjectivity, defeating the purpose of the approach.

Publication Bias: The File Drawer Problem

Another potential pitfall is the reliance on the available body of published studies, which may create exaggerated outcomes due to publication bias, as studies which show negative results or insignificant results are less likely to be published. For example, pharmaceutical companies have been known to hide negative studies and researchers may have overlooked unpublished studies such as dissertation studies or conference abstracts that did not reach publication. This is not easily solved, as one cannot know how many studies have gone unreported.

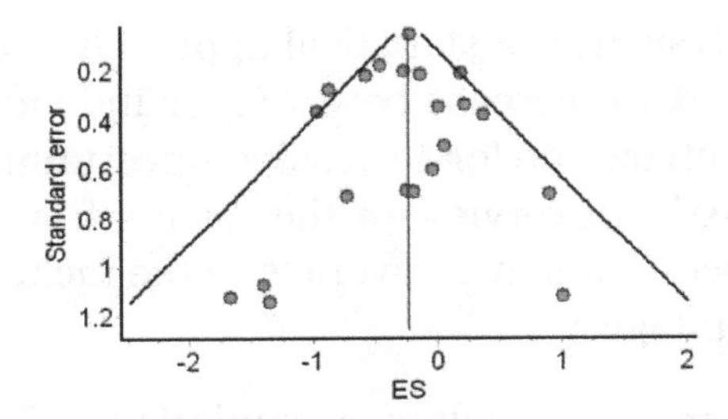

A funnel plot expected without the file drawer problem. The largest studies converge at the tip while smaller studies show more or less symmetrical scatter at the base

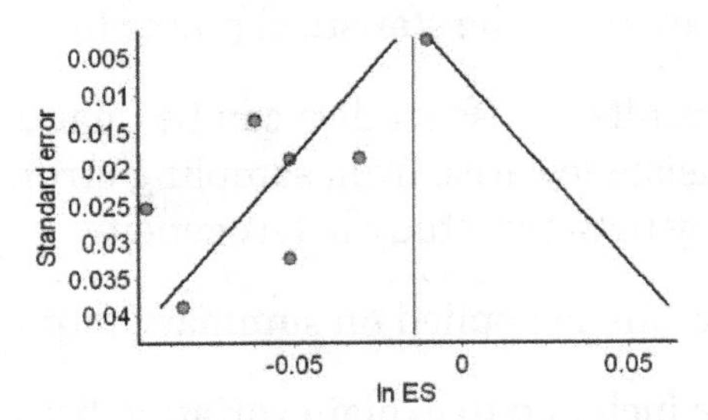

A funnel plot expected with the file drawer problem. The largest studies still cluster around the tip, but the bias against publishing negative studies has caused the smaller studies as a whole to have an unjustifiably favorable result to the hypothesis

This file drawer problem (characterized by negative or non-significant results being tucked away in a cabinet), can result in a biased distribution of effect sizes thus creating a serious base rate fallacy, in which the significance of the published studies is overestimated, as other studies were either not submitted for publication or were rejected. This should be seriously considered when interpreting the outcomes of a meta-analysis.

The distribution of effect sizes can be visualized with a funnel plot which (in its most common version) is a scatter plot of standard error versus the effect size. It makes use of the fact that the smaller studies (thus larger standard errors) have more scatter of the magnitude of effect (being less precise) while the larger studies have less scatter and form the tip of the funnel. If many negative studies were not published, the remaining positive studies give rise to a funnel plot in which the base is skewed to one side (asymmetry of the funnel plot). In contrast, when there is no publication bias, the effect of the smaller studies has no reason to be skewed to one side and so a symmetric funnel plot results. This also means that if no publication bias is present, there would be no relationship between standard error and effect size. A negative or positive relation between standard error and effect size would imply that smaller studies that found effects in one direction only were more likely to be published and/or to be submitted for publication.

Apart from the visual funnel plot, statistical methods for detecting publication bias have also been proposed. These are controversial because they typically have low power for detection of bias, but also may create false positives under some circumstances. For instance small study effects (biased smaller studies), wherein methodological differences between smaller and larger studies exist, may cause asymmetry in effect sizes that resembles publication bias. However, small study effects may be just as problematic for the interpretation of meta-analyses, and the imperative is on meta-analytic authors to investigate potential sources of bias.

A Tandem Method for analyzing publication bias has been suggested for cutting down false positive error problems. This Tandem method consists of three stages. Firstly, one calculates Orwin's fail-safe N, to check how many studies should be added in order to reduce the test statistic to a trivial size. If this number of studies is larger than the number of studies used in the meta-analysis, it is a sign that there is no publication bias, as in that case, one needs a lot of studies to reduce the effect size. Secondly, one can do an Egger's regression test, which tests whether the funnel plot is symmetrical. As mentioned before: a symmetrical funnel plot is a sign that there is no publication bias, as the effect size and sample size are not dependent. Thirdly, one can do the trim-and-fill method, which imputes data if the funnel plot is asymmetrical.

The problem of publication bias is not trivial as it is suggested that 25% of meta-analyses in the psychological sciences may have suffered from publication bias. However, low power of existing tests and problems with the visual appearance of the funnel plot remain an issue, and estimates of publication bias may remain lower than what truly exists.

Most discussions of publication bias focus on journal practices favoring publication of statistically significant findings. However, questionable research practices, such as reworking statistical models until significance is achieved, may also favor statistically significant findings in support of researchers' hypotheses.

Problems Related to the Statistical Approach

Other weaknesses are that it has not been determined if the statistically most accurate method for combining results is the fixed, IVhet, random or quality effect models, though the criticism against the random effects model is mounting because of the perception that the new random effects (used in meta-analysis) are essentially formal devices to facilitate smoothing or shrinkage and prediction may be impossible or ill-advised The main problem with the random effects approach is that it uses the classic statistical thought of generating a "compromise estimator" that makes the weights close to the naturally weighted estimator if heterogeneity across studies is large but close to the inverse variance weighted estimator if the between study heterogeneity is small. However, what has been ignored is the distinction between the model *we choose* to analyze a given dataset, and the *mechanism by which the data came into being*. A random effect can be present in either of these roles, but the two roles are quite distinct. There's no reason to think the analysis model and data-generation mechanism (model) are similar in form, but many sub-fields of statistics have developed the habit of assuming, for theory and simulations, that the data-generation mechanism (model) is identical to the analysis model we choose (or would like others to choose). As a hypothesized mechanisms for producing the data, the random effect model for meta-analysis is silly and it is more appropriate to think of this model as a superficial description and something we choose as an analytical tool – but this choice for meta-analysis may not work because the study effects are a fixed feature of the respective meta-analysis and the probability distribution is only a descriptive tool.

Problems Arising from Agenda-driven Bias

The most severe fault in meta-analysis often occurs when the person or persons doing the meta-analysis have an economic, social, or political agenda such as the passage or defeat of legislation. People with these types of agendas may be more likely to abuse meta-analysis due to personal bias. For example, researchers favorable to the author's agenda are likely to have their studies cherry-picked while those not favorable will be ignored or labeled as "not credible". In addition, the favored authors may themselves be biased or paid to produce results that support their overall political, social, or economic goals in ways such as selecting small favorable data sets and not incorporating larger unfavorable data sets. The influence of such biases on the results of a meta-analysis is possible because the methodology of meta-analysis is highly malleable.

A 2011 study done to disclose possible conflicts of interests in underlying research studies used for medical meta-analyses reviewed 29 meta-analyses and found that conflicts of interests in the studies underlying the meta-analyses were rarely disclosed. The 29 meta-analyses included 11 from general medicine journals, 15 from specialty medicine journals, and three from the Cochrane Database of Systematic Reviews. The 29 meta-analyses reviewed a total of 509 randomized controlled trials (RCTs). Of these, 318 RCTs reported funding sources, with 219 (69%) receiving funding from industry. Of the 509 RCTs, 132 reported author conflict of interest disclosures, with 91 studies (69%) disclosing one or more authors having financial ties to industry. The information was, however, seldom reflected in the meta-analyses. Only two (7%) reported RCT funding sources and none reported RCT author-industry ties. The authors concluded "without acknowledgment of COI due to industry funding or author industry financial ties from RCTs included in meta-analyses, readers' understanding and appraisal of the evidence from the meta-analysis may be compromised."

Methods and Assumptions

Approaches

In general, two types of evidence can be distinguished when performing a meta-analysis: Individual Participant Data (IPD), Aggregate Data (AD). The aggregate data can be direct or indirect.

AD is more commonly available (e.g. from the literature) and typically represents summary estimates such as odds ratios or relative risks. This can be directly synthesized across conceptually similar studies using several approaches. On the other hand, indirect aggregate data measures the effect of two treatments that were each compared against a similar control group in a meta-analysis. For example, if treatment A and treatment B were directly compared vs placebo in separate meta-analyses, we can use these two pooled results to get an estimate of the effects of A vs B in an indirect comparison as effect A vs Placebo minus effect B vs Placebo.

IPD evidence represents raw data as collected by the study centers. This distinction has raised the needs for different meta-analytic methods when evidence synthesis is desired, and has led to the development of one-stage and two-stage methods. In one-stage methods the IPD from all studies are modeled simultaneously whilst accounting for the clustering of participants within studies. Two-stage methods first compute summary statistics for AD from each study and then calculate overall statistics as a weighted average of the study statistics. By reducing IPD to AD, two-stage methods can also be applied when IPD is available; this makes them an appealing choice when performing a meta-analysis. Although it is conventionally believed that one-stage and two-stage methods yield similar results, recent studies have shown that they may occasionally lead to different conclusions.

Statistical Models for Aggregate Data

Direct Evidence: Models Incorporating Study Effects Only

Fixed Effects Model

The fixed effect model provides a weighted average of a series of study estimates. The inverse of the estimates' variance is commonly used as study weight, such that larger studies tend to contribute more than smaller studies to the weighted average. Consequently, when studies within a meta-analysis are dominated by a very large study, the findings from smaller studies are practically ignored. Most importantly, the fixed effects model assumes that all included studies investigate the same population, use the same variable and outcome definitions, etc. This assumption is typically unrealistic as research is often prone to several sources of heterogeneity; e.g. treatment effects may differ according to locale, dosage levels, study conditions, ...

Random effects model

A common model used to synthesize heterogeneous research is the random effects model of meta-analysis. This is simply the weighted average of the effect sizes of a group of studies. The weight that is applied in this process of weighted averaging with a random effects meta-analysis is achieved in two steps:

1. Step 1: Inverse variance weighting
2. Step 2: Un-weighting of this inverse variance weighting by applying a random effects variance component (REVC) that is simply derived from the extent of variability of the effect sizes of the underlying studies.

This means that the greater this variability in effect sizes (otherwise known as heterogeneity), the greater the un-weighting and this can reach a point when the random effects meta-analysis result becomes simply the un-weighted average effect size across the studies. At the other extreme, when all effect sizes are similar (or variability does not exceed sampling error), no REVC is applied and the random effects meta-analysis defaults to simply a fixed effect meta-analysis (only inverse variance weighting).

The extent of this reversal is solely dependent on two factors:

1. Heterogeneity of precision
2. Heterogeneity of effect size

Since neither of these factors automatically indicates a faulty larger study or more reliable smaller studies, the re-distribution of weights under this model will not bear a relationship to what these studies actually might offer. Indeed, it has been demonstrated that redistribution of weights is simply in one direction from larger to smaller studies as heterogeneity increases until eventually all studies have equal weight and no more redistribution is possible. Another issue with the random effects model is that the most commonly used confidence intervals generally do not retain their coverage probability above the specified nominal level and thus substantially underestimate the statistical error and are potentially overconfident in their conclusions. Several fixes have been suggested but the debate continues on. A further concern is that the average treatment effect can sometimes be even less conservative compared to the fixed effect model and therefore misleading in practice. One interpretational fix that has been suggested is to create a prediction interval around the random effects estimate to portray the range of possible effects in practice. However, an assumption behind the calculation of such a prediction interval is that trials are considered more or less homogeneous entities and that included patient populations and comparator treatments should be considered exchangeable and this is usually unattainable in practice.

The most widely used method to estimate between studies variance (REVC) is the DerSimonian-Laird (DL) approach. Several advanced iterative (and computationally expensive) techniques for computing the between studies variance exist (such as maximum likelihood, profile likelihood and restricted maximum likelihood methods) and random effects models using these methods can be run in Stata with the metaan command. The metaan command must be distinguished from the classic metan (single "a") command in Stata that uses the DL estimator. These advanced methods have also been implemented in a free and easy to use Microsoft Excel add-on, MetaEasy. However, a comparison between these advanced methods and the DL method of computing the between studies variance demonstrated that there is little to gain and DL is quite adequate in most scenarios.

However, most meta-analyses include between 2-4 studies and such a sample is more often than not inadequate to accurately estimate heterogeneity. Thus it appears that in small meta-analyses, an incorrect zero between study variance estimate is obtained, leading to a false homogeneity assumption. Overall, it appears that heterogeneity is being consistently underestimated in meta-analyses and sensitivity analyses in which high heterogeneity levels are assumed could be informative. These random effects models and software packages mentioned above relate to study-aggregate meta-analyses and researchers wishing to conduct individual patient data (IPD) meta-analyses need to consider mixed-effects modelling approaches.

IVhet model

Doi & Barendregt working in collaboration with Khan, Thalib and Williams (from the University of Queensland, University of Southern Queensland and Kuwait University), have created an inverse variance quasi likelihood based alternative (IVhet) to the random effects (RE) model for which details are available online. This was incorporated into MetaXL version 2.0, a free Microsoft excel add-in for meta-analysis produced by Epigear International Pty Ltd, and made available on 5 April 2014. The authors state that a clear advantage of this model is that it resolves the two main problems of the random effects model. The first advantage of the IVhet model is that coverage remains at the nominal (usually 95%) level for the confidence interval unlike the random effects model which drops in coverage with increasing heterogeneity. The second advantage is that the IVhet model maintains the inverse variance weights of individual studies, unlike the RE model which gives small studies more weight (and therefore larger studies less) with increasing heterogeneity. When heterogeneity becomes large, the individual study weights under the RE model become equal and thus the RE model returns an arithmetic mean rather than a weighted average. This side-effect of the RE model does not occur with the IVhet model which thus differs from the RE model estimate in two perspectives: Pooled estimates will favor larger trials (as opposed to penalizing larger trials in the RE model) and will have a confidence interval that remains within the nominal coverage under uncertainty (heterogeneity). Doi & Barendregt suggest that while the RE model provides an alternative method of pooling the study data, their simulation results demonstrate that using a more specified probability model with untenable assumptions, as with the RE model, does not necessarily provide better results. Tha latter study also reports that the IVhet model resolves the problems related to underestimation of the statistical error, poor coverage of the confidence interval and increased MSE seen with the random effects model and the authors conclude that researchers should henceforth abandon use of the random effects model in meta-analysis. While their data is compelling, the ramifications (in terms of the magnitude of spuriously positive results within the Cochrane database) are huge and thus accepting this conclusion requires careful independent confirmation. The availability of a free software (MetaXL) that runs the IVhet model (and all other models for comparison) facilitates this for the research community.

Direct Evidence: Models Incorporating Additional Information

Quality Effects Model

Doi and Thalib originally introduced the quality effects model. They introduce a new approach to adjustment for inter-study variability by incorporating the contribution of variance due to a relevant component (quality) in addition to the contribution of variance due to random error that is used in any fixed effects meta-analysis model to generate weights for each study. The strength of the quality effects meta-analysis is that it allows available methodological evidence to be used over subjective random effects, and thereby helps to close the damaging gap which has opened up between methodology and statistics in clinical research. To do this a synthetic bias variance is computed based on quality information to adjust inverse variance weights and the quality adjusted weight of the ith study is introduced. These adjusted weights are then used in meta-analysis. In other words, if study i is of good quality and other studies are of poor quality, a proportion of their quality adjusted weights is mathematically redistributed to study i giving it more weight towards the overall effect size. As studies become increasingly similar in terms of quality, re-distribution becomes progressively less and ceases when all studies are of equal quality (in the case of equal quality, the quality effects model defaults to the IVhet model). A recent evaluation of the quality effects model (with some updates) demonstrates that despite the subjectivity of quality assessment, the performance (MSE and true variance under simulation) is superior to that achievable with the random effects model. This model thus replaces the untenable interpretations that abound in the literature and a software is available to explore this method further.

Indirect Evidence: Network Meta-analysis Methods

Indirect comparison meta-analysis methods (also called network meta-analyses, in particular when multiple treatments are assessed simultaneously) generally use two main methodologies. First, is the Bucher method which is a single or repeated comparison of a closed loop of three-treatments such that one of them is common to the two studies and forms the node where the loop begins and ends. Therefore, multiple two-by-two comparisons (3-treatment loops) are needed to compare multiple treatments. This methodology requires that trials with more than two arms have two arms only selected as independent pair-wise comparisons are required. The alternative methodology uses complex statistical modelling to include the multiple arm trials and comparisons simultaneously between all competing treatments. These have been executed using Bayesian methods, mixed linear models and meta-regression approaches

Bayesian framework

Specifying a Bayesian network meta-analysis model involves writing a directed acyclic graph (DAG) model for general-purpose Markov chain Monte Carlo (MCMC) software such as WinBUGS. In addition, prior distributions have to be specified for a number of the parameters, and the data have to be supplied in a specific format. Together, the DAG, priors, and data form a Bayesian hierarchical model. To complicate matters further, because of the nature of MCMC estimation, overdispersed starting values have to be chosen for a number of independent chains so that convergence can be assessed. Currently, there is no software that automatically generates such models, although there are some tools to aid in the process. The complexity of the Bayesian approache has limited usage of this methodology. Methodology for automation of this method has been suggested but requires that arm-level outcome data are available, and this is usually unavailable. Great claims are sometimes made for the inherent ability of the Bayesian framework to handle network meta-analysis and its greater flexibility. However, this choice of implementation of framework for inference, Bayesian or frequentist, may be less important than other choices regarding the modeling of effects.

Frequentist multivariate framework

On the other hand, the frequentist multivariate methods involve approximations and assumptions that are not stated explicitly or verified when the methods are applied. For example, The mvmeta package for Stata enables network meta-analysis in a frequentist framework. However, if there is no common comparator in the network, then this has to be handled by augmenting the dataset with fictional arms with high variance, which is not very objective and requires a decision as to what constitutes a sufficiently high variance. The other issue is use of the random effects model in both this frequentist framework and the Bayesian framework. Senn advises analysts to be cautious about interpreting the 'random effects' analysis since only one random effect is allowed for but one could envisage many. Senn goes on to say that it is rather naıve, even in the case where only two treatments are being compared to assume that random-effects analysis accounts for all uncertainty about the way effects can vary from trial to trial. Newer models of meta-analysis such as those discussed above would certainly help alleviate this situation and have been implemented in the next framework.

Generalized pairwise modelling framework

An approach that has been tried since the late 1990s is the implementation of the multiple three-treatment closed-loop analysis. This has not been popular because the process rapidly becomes overwhelming as network complexity increases. Development in this area was then abandoned in favor of the Bayesian and multivariate frequentist methods which emerged as alternatives. Very recently, automation of the three-treatment closed loop method has been developed for complex networks by some researchers as a way to make this methodology available to the mainstream research community. This proposal does restrict each trial to two interventions, but also introduces a workaround for multiple arm trials: a different fixed control node can be selected in different runs. It also utilizes robust meta-analysis methods so that many of the problems highlighted above are avoided. Further research around this framework is required to determine if this is indeed superior to the Bayesian or multivariate frequentist frameworks. Researchers willing to try this out have access to this framework through a free software.

Applications in Modern Science

Modern statistical meta-analysis does more than just combine the effect sizes of a set of studies using a weighted average. It can test if the outcomes of studies show more variation than the variation that is expected because of the sampling of different numbers of research participants. Additionally, study characteristics such as measurement instrument used, population sampled, or aspects of the studies' design can be coded and used to reduce variance of the estimator. Thus some methodological weaknesses in studies can be corrected statistically. Other uses of meta-analytic methods include the development of clinical prediction models, where meta-analysis may be used to combine data from different research centers, or even to aggregate existing prediction models.

Meta-analysis can be done with single-subject design as well as group research designs. This is important because much research has been done with single-subject research designs. Considerable dispute exists for the most appropriate meta-analytic technique for single subject research.

Meta-analysis leads to a shift of emphasis from single studies to multiple studies. It emphasizes the practical importance of the effect size instead of the statistical significance of individual studies. This shift in thinking has been termed "meta-analytic thinking". The results of a meta-analysis are often shown in a forest plot.

Results from studies are combined using different approaches. One approach frequently used in meta-analysis in health care research is termed 'inverse variance method'. The average effect size across all studies is computed as a *weighted mean,* whereby the weights are equal to the inverse variance of each studies' effect estimator. Larger studies and studies with less random variation are given greater weight than smaller studies. Other common approaches include the Mantel–Haenszel method and the Peto method.

A recent approach to studying the influence that weighting schemes can have on results has been proposed through the construct of *gravity,* which is a special case of combinatorial meta-analysis.

Signed differential mapping is a statistical technique for meta-analyzing studies on differences in brain activity or structure which used neuroimaging techniques such as fMRI, VBM or PET.

Different high throughput techniques such as microarrays have been used to understand Gene expression. MicroRNA expression profiles have been used to identify differentially expressed microRNAs in particular cell or tissue type or disease conditions or to check the effect of a treatment. A meta-analysis of such expression profiles was performed to derive novel conclusions and to validate the known findings.=

References

- Stack, Carol B. (1974). All Our Kin:Strategies for Survival in a black community. New York, New York: Harper and Row. ISBN 0-06-013974-9.
- Olaf Zenker & Karsten Kumoll. Beyond Writing Culture: Current Intersections of Epistemologies and Representational Practices. (2010). New York: Berghahn Books. ISBN 978-1-84545-675-7.
- Paul A. Erickson & Liam D. Murphy. A History of Anthropological Theory, Third Edition. (2008). Toronto: Broadview Press. ISBN 978-1-55111-871-0. Pg. 190
- Olaf Zenker & Karsten Kumoll. Beyond Writing Culture: Current Intersections of Epistemologies and Representational Practices. (2010). New York: Berghahn Books. ISBN 978-1-84545-675-7. Pg. 12
- Savin-Baden, M.; Major, C. (2013). Qualitative Research: The Essential Guide to Theory and Practice. London and New York: Routledge. ISBN 978-0-415-67478-2.
- Aldiabat, Khaldoun; Navenec, Carole-Lynne (4 July 2011). "Philosophical Roots of Classical Grounded Theory: Its Foundations in Symbolic Interactionism" (PDF). The Qualitative Report. 16: 1063–80. Retrieved 5 December 2014.
- "Clarification of the Blurred Boundaries between Grounded Theory and Ethnography: Differences and Similarities" (PDF). Turkish Online Journal of Qualitative Inquiry. 2. July 2011. Retrieved 5 December 2014.

5

Fundamentals of Research

The important aspects of research are hypothesis, theory, law, conceptual model and generalization. Hypothesis is an explanation that is suggested for a particular phenomenon whereas theory is an abstraction which is rational in its approach and tries to explain thought. The chapter serves as a source to understand the major classifications related to the fundamentals of research.

Hypothesis

A hypothesis (plural hypotheses) is a proposed explanation for a phenomenon. For a hypothesis to be a scientific hypothesis, the scientific method requires that one can test it. Scientists generally base scientific hypotheses on previous observations that cannot satisfactorily be explained with the available scientific theories. Even though the words "hypothesis" and "theory" are often used synonymously, a scientific hypothesis is not the same as a scientific theory. A working hypothesis is a provisionally accepted hypothesis proposed for further research.

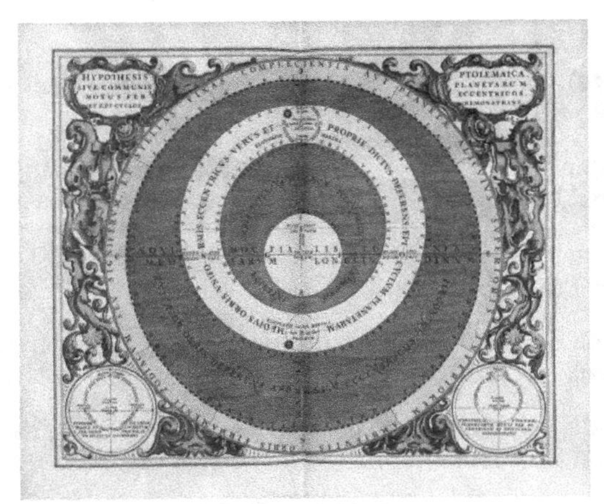

The hypothesis of Andreas Cellarius, showing the planetary motions in eccentric and epicyclical orbits.

A different meaning of the term *hypothesis* is used in formal logic, to denote the antecedent of a proposition; thus in the proposition "If *P*, then *Q*", *P* denotes the hypothesis (or antecedent); *Q* can be called a consequent. *P* is the assumption in a (possibly counterfactual) *What If* question.

The adjective *hypothetical*, meaning "having the nature of a hypothesis", or "being assumed to exist as an immediate consequence of a hypothesis", can refer to any of these meanings of the term "hypothesis".

Uses

Remember, the way that you prove an implication is by assuming the hypothesis. --Philip Wadler

In its ancient usage, *hypothesis* referred to a summary of the plot of a classical drama. The English word *hypothesis* comes from the ancient Greek word *hupothesis*, meaning "to put under" or "to suppose".

In Plato's *Meno* (86e–87b), Socrates dissects virtue with a method used by mathematicians, that of "investigating from a hypothesis." In this sense, 'hypothesis' refers to a clever idea or to a convenient mathematical approach that simplifies cumbersome calculations. Cardinal Bellarmine gave a famous example of this usage in the warning issued to Galileo in the early 17th century: that he must not treat the motion of the Earth as a reality, but merely as a hypothesis.

In common usage in the 21st century, a *hypothesis* refers to a provisional idea whose merit requires evaluation. For proper evaluation, the framer of a hypothesis needs to define specifics in operational terms. A hypothesis requires more work by the researcher in order to either confirm or disprove it. In due course, a confirmed hypothesis may become part of a theory or occasionally may grow to become a theory itself. Normally, scientific hypotheses have the form of a mathematical model. Sometimes, but not always, one can also formulate them as existential statements, stating that some particular instance of the phenomenon under examination has some characteristic and causal explanations, which have the general form of universal statements, stating that every instance of the phenomenon has a particular characteristic.

In Entrepreneurial science, a hypothesis is used to formulate provisional ideas within a business setting. The formulated hypothesis is then evaluated where either the hypothesis is proven to be "true" or "false" through a verifiability- or falsifiability-oriented Experiment.

Any useful hypothesis will enable predictions by reasoning (including deductive reasoning). It might predict the outcome of an experiment in a laboratory setting or the observation of a phenomenon in nature. The prediction may also invoke statistics and only talk about probabilities. Karl Popper, following others, has argued that a hypothesis must be falsifiable, and that one cannot regard a proposition or theory as scientific if it does not admit the possibility of being shown false. Other philosophers of science have rejected the criterion of falsifiability or supplemented it with other criteria, such as verifiability (e.g., verificationism) or coherence (e.g., confirmation holism). The scientific method involves experimentation, to test the ability of some hypothesis to adequately answer the question under investigation. In contrast, unfettered observation is not as likely to raise unexplained issues or open questions in science, as would the formulation of a crucial experiment to test the hypothesis. A thought experiment might also be used to test the hypothesis as well.

In framing a hypothesis, the investigator must not currently know the outcome of a test or that it remains reasonably under continuing investigation. Only in such cases does the experiment, test or study potentially increase the probability of showing the truth of a hypothesis. If the researcher already knows the outcome, it counts as a "consequence" — and the researcher should have already considered this while formulating the hypothesis. If one cannot assess the predictions by observation or by experience, the hypothesis needs to be tested by others providing observations. For example, a new technology or theory might make the necessary experiments feasible.

Scientific Hypothesis

People refer to a trial solution to a problem as a hypothesis, often called an "educated guess" because it provides a suggested solution based on the evidence. However, some scientists reject the term "educated guess" as incorrect. Experimenters may test and reject several hypotheses before solving the problem.

According to Schick and Vaughn, researchers weighing up alternative hypotheses may take into consideration:

- Testability (compare falsifiability as discussed above)
- Parsimony (as in the application of "Occam's razor", discouraging the postulation of excessive numbers of entities)
- Scope – the apparent application of the hypothesis to multiple cases of phenomena
- Fruitfulness – the prospect that a hypothesis may explain further phenomena in the future
- Conservatism – the degree of "fit" with existing recognized knowledge-systems.

Working Hypothesis

A working hypothesis is a hypothesis that is provisionally accepted as a basis for further research in the hope that a tenable theory will be produced, even if the hypothesis ultimately fails. Like all hypotheses, a working hypothesis is constructed as a statement of expectations, which can be linked to the exploratory research purpose in empirical investigation. Working hypotheses are often used as a conceptual framework in qualitative research.

The provisional nature of working hypotheses make them useful as an organizing device in applied research. Here they act like a useful guide to address problems that are still in a formative phase.

In recent years, philosophers of science have tried to integrate the various approaches to evaluating hypotheses, and the scientific method in general, to form a more complete system that integrates the individual concerns of each approach. Notably, Imre Lakatos and Paul Feyerabend, Karl Popper's colleague and student, respectively, have produced novel attempts at such a synthesis.

Hypotheses, Concepts and Measurement

Concepts in Hempel's deductive-nomological model play a key role in the development and testing of hypotheses. Most formal hypotheses connect concepts by specifying the expected relationships between propositions. When a set of hypotheses are grouped together they become a type of conceptual framework. When a conceptual framework is complex and incorporates causality or explanation it is generally referred to as a theory. According to noted philosopher of science Carl Gustav Hempel "An adequate empirical interpretation turns a theoretical system into a testable theory: The hypothesis whose constituent terms have been interpreted become capable of test by reference to observable phenomena. Frequently the interpreted hypothesis will be derivative hypotheses of the theory; but their confirmation or disconfirmation by empirical data will then immediately strengthen or weaken also the primitive hypotheses from which they were derived."

Hempel provides a useful metaphor that describes the relationship between a conceptual framework and the framework as it is observed and perhaps tested (interpreted framework). "The whole system floats, as it were, above the plane of observation and is anchored to it by rules of interpretation. These might be viewed as strings which are not part of the network but link certain points of the latter with specific places in the plane of observation. By virtue of those interpretative connections, the network can function as a scientific theory" Hypotheses with concepts anchored in the plane of observation are ready to be tested. In "actual scientific practice the process of framing a theoretical structure and of interpreting it are not always sharply separated, since the intended interpretation usually guides the construction of the theoretician." It is, however, "possible and indeed desirable, for the purposes of logical clarification, to separate the two steps conceptually."

Statistical Hypothesis Testing

When a possible correlation or similar relation between phenomena is investigated, such as whether a proposed remedy is effective in treating a disease, the hypothesis that a relation exists cannot be examined the same way one might examine a proposed new law of nature. In such an investigation, if the tested remedy shows no effect in a few cases, these do not necessarily falsify the hypothesis. Instead, statistical tests are used to determine how likely it is that the overall effect would be observed if the hypothesized relation does not exist. If that likelihood is sufficiently small (e.g., less than 1%), the existence of a relation may be assumed. Otherwise, any observed effect may be due to pure chance.

In statistical hypothesis testing, two hypotheses are compared. These are called the null hypothesis and the alternative hypothesis. The null hypothesis is the hypothesis that states that there is no relation between the phenomena whose relation is under investigation, or at least not of the form given by the alternative hypothesis. The alternative hypothesis, as the name suggests, is the alternative to the null hypothesis: it states that there *is* some kind of relation. The alternative hypothesis may take several forms, depending on the nature of the hypothesized relation; in particular, it can be two-sided (for example: there is *some* effect, in a yet unknown direction) or one-sided (the direction of the hypothesized relation, positive or negative, is fixed in advance).

Conventional significance levels for testing hypotheses (acceptable probabilities of wrongly rejecting a true null hypothesis) are .10, .05, and .01. Whether the null hypothesis is rejected and the alternative hypothesis is accepted, must be determined in advance, before the observations are collected or inspected. If these criteria are determined later, when the data to be tested are already known, the test is invalid.

The above procedure is actually dependent on the number of the participants (units or sample size) that is included in the study. For instance, the sample size may be too small to reject a null hypothesis and, therefore, it is recommended to specify the sample size from the beginning. It is advisable to define a small, medium and large effect size for each of a number of important statistical tests which are used to test the hypotheses.

Theory

Theory is a contemplative and rational type of abstract or generalizing thinking, or the results of such thinking. Depending on the context, the results might for example include generalized explanations of how nature works. The word has its roots in ancient Greek, but in modern use it has taken on several different related meanings.

A theory can be *normative* (or prescriptive), meaning a postulation about what ought to be. It provides "goals, norms, and standards". A theory can be a body of knowledge, which may or may not be associated with particular explanatory models. To theorize is to develop this body of knowledge.

As already in Aristotle's definitions, theory is very often contrasted to "practice" a Greek term for "doing", which is opposed to theory because pure theory involves no doing apart from itself. A classical example of the distinction between "theoretical" and "practical" uses the discipline of medicine: medical theory involves trying to understand the causes and nature of health and sickness, while the practical side of medicine is trying to make people healthy. These two things are related but can be independent, because it is possible to research health and sickness without curing specific patients, and it is possible to cure a patient without knowing how the cure worked.

In modern science, the term "theory" refers to scientific theories, a well-confirmed type of explanation of nature, made in a way consistent with scientific method, and fulfilling the criteria required by modern science. Such theories are described in such a way that any scientist in the field is in a position to understand and either provide empirical support ("verify") or empirically contradict ("falsify") it. Scientific theories are the most reliable, rigorous, and comprehensive form of scientific knowledge, in contrast to more common uses of the word "theory" that imply that something is unproven or speculative (which is better characterized by the word 'hypothesis'). Scientific theories are distinguished from hypotheses, which are individual empirically testable conjectures, and scientific laws, which are descriptive accounts of how nature will behave under certain conditions.

Ancient Uses

The English word *theory* was derived from a technical term in philosophy in Ancient Greek. As an everyday word, *theoria*, meant "a looking at, viewing, beholding", but in more technical contexts it came to refer to contemplative or speculative understandings of natural things, such as those of natural philosophers, as opposed to more practical ways of knowing things, like that of skilled orators or artisans. The word has been in use in English since at least the late 16th century. Modern uses of the word "theory" are derived from the original definition, but have taken on new shades of meaning, still based on the idea that a theory is a thoughtful and rational explanation of the general nature of things.

In the book *From Religion to Philosophy*, Francis Cornford suggests that the Orphics used the word "theory" to mean 'passionate sympathetic contemplation'. Pythagoras changed the word to mean a passionate sympathetic contemplation of mathematical knowledge, because he considered this intellectual pursuit the way to reach the highest plane of existence. Pythagoras emphasized subduing emotions and bodily desires in order to enable the intellect to function at the higher plane of theory. Thus it was Pythagoras who gave the word "theory" the specific meaning which leads to the classical and modern concept of a distinction between theory as uninvolved, neutral thinking, and practice.

In Aristotle's terminology, as has already been mentioned above, theory is contrasted with *praxis* or practice, which remains the case today. For Aristotle, both practice and theory involve thinking, but the aims are different. Theoretical contemplation considers things which humans do not move or change, such as nature, so it has no human aim apart from itself and the knowledge it helps create. On the other hand, *praxis* involves thinking, but always with an aim to desired actions, whereby humans cause change or movement themselves for their own ends. Any human movement which involves no conscious choice and thinking could not be an example of *praxis* or doing.

Theories Formally and Scientifically

Theories are analytical tools for understanding, explaining, and making predictions about a given subject matter. There are theories in many and varied fields of study, including the arts and sciences. A formal theory is syntactic in nature and is only meaningful when given a semantic component by applying it to some content (e.g., facts and relationships of the actual historical world as it is unfolding). Theories in various fields of study are expressed in natural language, but are always constructed in such a way that their general form is identical to a theory as it is expressed in the formal language of mathematical logic. Theories may be expressed mathematically, symbolically, or in common language, but are generally expected to follow principles of rational thought or logic.

Theory is constructed of a set of sentences which consists entirely of true statements about the subject matter under consideration. However, the truth of any one of these statements is always relative to the whole theory. Therefore, the same statement may be true with respect to one theory, and not true with respect to another. This is, in ordinary language, where statements such as "He is a terrible person" cannot be judged to be true or false without reference to some interpretation of who "He" is and for that matter what a "terrible person" is under the theory.

Sometimes two theories have exactly the same explanatory power because they make the same predictions. A pair of such theories is called indistinguishable or observationally equivalent, and the choice between them reduces to convenience or philosophical preference.

The form of theories is studied formally in mathematical logic, especially in model theory. When theories are studied in mathematics, they are usually expressed in some formal language and their statements are closed under application of certain procedures called rules of inference. A special case of this, an axiomatic theory, consists of axioms (or axiom schemata) and rules of inference. A theorem is a statement that can be derived from those axioms by application of these rules of inference. Theories used in applications are abstractions of observed phenomena and the resulting theorems provide solutions to real-world problems. Obvious examples include arithmetic (abstracting concepts of number), geometry (concepts of space), and probability (concepts of randomness and likelihood).

Gödel's incompleteness theorem shows that no consistent, recursively enumerable theory (that is, one whose theorems form a recursively enumerable set) in which the concept of natural numbers can be expressed, can include all true statements about them. As a result, some domains of knowledge cannot be formalized, accurately and completely, as mathematical theories. (Here, formalizing accurately and completely means that all true propositions—and only true propositions—are derivable within the mathematical system.) This limitation, however, in no way precludes the construction of mathematical theories that formalize large bodies of scientific knowledge.

Underdetermination

A theory is *underdetermined* (also called *indeterminacy of data to theory*) if a rival, inconsistent theory is at least as consistent with the evidence. Underdetermination is an epistemological issue about the relation of evidence to conclusions.

A theory that lacks supporting evidence is generally, more properly, referred to as a hypothesis.

Intertheoretic Reduction and Elimination

If there is a new theory which is better at explaining and predicting phenomena than an older theory (i.e. it has more explanatory power), we are justified in believing that the newer theory describes reality more correctly. This is called an *intertheoretic reduction* because the terms of the old theory can be reduced to the terms of the new one. For instance, our historical understanding about "sound", "light" and "heat" have today been reduced to "wave compressions and rarefactions", "electromagnetic waves", and "molecular kinetic energy", respectively. These terms which are identified with each other are called *intertheoretic identities*. When an old theory and a new one are parallel in this way, we can conclude that we are describing the same reality, only more completely.

In cases where a new theory uses new terms which do not reduce to terms of an older one, but rather replace them entirely because they are actually a misrepresentation it is called an *intertheoretic elimination*. For instance, the obsolete scientific theory that put forward an understanding of heat transfer in terms of the movement of caloric fluid was eliminated when a theory of heat as energy replaced it. Also, the theory that phlogiston is a substance released from burning and rusting material was eliminated with the new understanding of the reactivity of oxygen.

Theories Vs. Theorems

Theories are distinct from theorems. Theorems are derived deductively from objections according to a formal system of rules, sometimes as an end in itself and sometimes as a first step in testing or applying a theory in a concrete situation; theorems are said to be true in the sense that the conclusions of a theorem are logical consequences of the objections. Theories are abstract and conceptual, and to this end they are always considered true. They are supported or challenged by observations in the world. They are 'rigorously tentative', meaning that they are proposed as true and expected to satisfy careful examination to account for the possibility of faulty inference or incorrect observation. Sometimes theories are incorrect, meaning that an explicit set of observations contradicts some fundamental objection or application of the theory, but more often theories are corrected to conform to new observations, by restricting the class of phenomena the theory applies to or changing the assertions made. An example of the former is the restriction of Classical mechanics to phenomena involving macroscopic length scales and particle speeds much lower than the speed of light.

"Sometimes a hypothesis never reaches the point of being considered a theory because the answer is not found to derive its assertions analytically or not applied empirically."

Philosophical Theories

Theories whose subject matter consists not in empirical data, but rather in ideas are in the realm of *philosophical theories* as contrasted with *scientific theories*. At least some of the elementary theorems of a philosophical theory are statements whose truth cannot necessarily be scientifically tested through empirical observation.

Fields of study are sometimes named "theory" because their basis is some initial set of objections describing the field's approach to a subject matter. These assumptions are the elementary theorems of the particular theory, and can be thought of as the axioms of that field. Some commonly known examples include set theory and number theory; however literary theory, critical theory, and music theory are also of the same form.

Metatheory

One form of philosophical theory is a *metatheory* or *meta-theory*. A metatheory is a theory whose subject matter is some other theory. In other words, it is a theory about a theory. Statements made in the metatheory about the theory are called metatheorems.

Political Theories

A political theory is an ethical theory about the law and government. Often the term "political theory" refers to a general view, or specific ethic, political belief or attitude, about politics.

Scientific Theories

In science, the term "theory" refers to "a well-substantiated explanation of some aspect of the natural world, based on a body of facts that have been repeatedly confirmed through observation and experiment." Theories must also meet further requirements, such as the ability to make falsifiable predictions with consistent accuracy across a broad area of scientific inquiry, and production of strong evidence in favor of the theory from multiple independent sources.

The strength of a scientific theory is related to the diversity of phenomena it can explain, which is measured by its ability to make falsifiable predictions with respect to those phenomena. Theories are improved (or replaced by better theories) as more evidence is gathered, so that accuracy in prediction improves over time; this increased accuracy corresponds to an increase in scientific knowledge. Scientists use theories as a foundation to gain further scientific knowledge, as well as to accomplish goals such as inventing technology or curing disease.

Definitions from Scientific Organizations

The United States National Academy of Sciences defines scientific theories as follows:

The formal scientific definition of "theory" is quite different from the everyday meaning of the word. It refers to a comprehensive explanation of some aspect of nature that is supported by a vast body of evidence. Many scientific theories are so well established that no new evidence is likely to alter them substantially. For example, no new evidence will demonstrate that the Earth does not orbit around the sun (heliocentric theory), or that living things are not made of cells (cell theory), that matter is not composed of atoms, or that the surface of the Earth is not divided into solid plates that have moved over geological timescales (the theory of plate tectonics)...One of the most useful properties of scientific theories is that they can be used to make predictions about natural events or phenomena that have not yet been observed.

From the American Association for the Advancement of Science:

A scientific theory is a well-substantiated explanation of some aspect of the natural world, based on a body of facts that have been repeatedly confirmed through observation and experiment. Such fact-supported theories are not "guesses" but reliable accounts of the real world. The theory of biological evolution is more than "just a theory." It is as factual an explanation of the universe as the atomic theory of matter or the germ theory of disease. Our understanding of gravity is still a work in progress. But the phenomenon of gravity, like evolution, is an accepted fact.

Note that the term *theory* would not be appropriate for describing untested but intricate hypotheses or even scientific models.

Philosophical Views

The logical positivists thought of scientific theories as *deductive theories* — that a theory's content is based on some formal system of logic and on basic axioms. In a deductive theory, any sentence which is a logical consequence of one or more of the axioms is also a sentence of that theory. This is called the received view of theories.

In the semantic view of theories, which has largely replaced the received view, theories are viewed as scientific models. A model is a logical framework intended to represent reality (a "model of reality"), similar to the way that a map is a graphical model that represents the territory of a city or country. In this approach, theories are a specific category of models which fulfill the necessary criteria.

In Physics

In physics the term *theory* is generally used for a mathematical framework—derived from a small set of basic postulates (usually symmetries, like equality of locations in space or in time, or identity of electrons, etc.)—which is capable of producing experimental predictions for a given category of physical systems. One good example is classical electromagnetism, which encompasses results derived from gauge symmetry (sometimes called gauge invariance) in a form of a few equations called Maxwell's equations. The specific mathematical aspects of classical electromagnetic theory are termed "laws of electromagnetism", reflecting the level of consistent and reproducible evidence that supports them. Within electromagnetic theory generally, there are numerous hypotheses about how electromagnetism applies to specific situations. Many of these hypotheses are already considered to be adequately tested, with new ones always in the making and perhaps untested.

The Term Theoretical

Acceptance of a theory does not require that all of its major predictions be tested, if it is already supported by sufficiently strong evidence. For example, certain tests may be unfeasible or technically difficult. As a result, theories may make predictions that have not yet been confirmed or proven incorrect; in this case, the predicted results may be described informally with the term "theoretical." These predictions can be tested at a later time, and if they are incorrect, this may lead to revision or rejection of the theory.

List of Notable Theories

Most of the following are scientific theories; some are not, but rather encompass a body of knowledge or art, such as Music theory and Visual Arts Theories.

- Anthropology: Carneiro's circumscription theory
- Astronomy: Alpher–Bethe–Gamow theory — B²FH Theory — Copernican theory — Giant impact hypothesis — Newton's theory of gravitation — Hubble's Law — Kepler's laws of planetary motion — Nebular hypothesis — Ptolemaic theory
- Cosmology: Big Bang Theory — Cosmic inflation — Loop quantum gravity — Superstring theory — Supergravity — Supersymmetric theory — Multiverse theory — Holographic principle — Quantum gravity — M-theory

- Biology: Cell theory — Evolution — Germ theory
- Chemistry: Molecular theory — Kinetic theory of gases — Molecular orbital theory — Valence bond theory — Transition state theory — RRKM theory — Chemical graph theory — Flory–Huggins solution theory — Marcus theory — Lewis theory (successor to Brønsted–Lowry acid–base theory) — HSAB theory — Debye–Hückel theory — Thermodynamic theory of polymer elasticity — Reptation theory — Polymer field theory — Møller–Plesset perturbation theory — density functional theory — Frontier molecular orbital theory — Polyhedral skeletal electron pair theory — Baeyer strain theory — Quantum theory of atoms in molecules — Collision theory — Ligand field theory (successor to Crystal field theory) — Variational Transition State Theory — Benson group increment theory — Specific ion interaction theory
- Climatology: Climate change theory (general study of climate changes) and anthropogenic climate change (ACC)/ global warming (AGW) theories (due to human activity)
- Economics: Macroeconomic theory — Microeconomic theory — Law of Supply and demand
- Education: Constructivist theory — Critical pedagogy theory — Education theory — Multiple intelligence theory — Progressive education theory
- Engineering: Circuit theory — Control theory — Signal theory — Systems theory — Information theory
- Film: Film Theory
- Geology: Plate tectonics
- Humanities: Critical theory
- Linguistics: X-bar theory — Government and Binding — Principles and parameters — Universal grammar
- Literature: Literary theory
- Mathematics: Approximation theory — Arakelov theory — Asymptotic theory — Bifurcation theory — Catastrophe theory — Category theory — Chaos theory — Choquet theory — Coding theory — Combinatorial game theory — Computability theory — Computational complexity theory — Deformation theory — Dimension theory — Ergodic theory — Field theory — Galois theory — Game theory — Graph theory — Group theory — Hodge theory — Homology theory — Homotopy theory — Ideal theory — Intersection theory — Invariant theory — Iwasawa theory — K-theory — KK-theory — Knot theory — L-theory — Lie theory — Littlewood–Paley theory — Matrix theory — Measure theory — Model

theory — Morse theory — Nevanlinna theory — Number theory — Obstruction theory — Operator theory — PCF theory — Perturbation theory — Potential theory — Probability theory — Ramsey theory — Rational choice theory — Representation theory — Ring theory — Set theory — Shape theory — Small cancellation theory — Spectral theory — Stability theory — Stable theory — Sturm–Liouville theory — Twistor theory

- Music: Music theory
- Philosophy: Proof theory — Speculative reason — Theory of truth — Type theory — Value theory — Virtue theory
- Physics: Acoustic theory — Antenna theory — Atomic theory — BCS theory — Dirac hole theory — Dynamo theory — Landau theory — M-theory — Perturbation theory — Theory of relativity (successor to classical mechanics) — Quantum field theory — Scattering theory — String theory — Quantum information theory
- Psychology: Theory of mind — Cognitive dissonance theory — Attachment theory — Object permanence — Poverty of stimulus — Attribution theory — Self-fulfilling prophecy — Stockholm syndrome
- Semiotics: Intertheoricity - Transferogenesis
- Sociology: Critical theory — Engaged theory — Social theory — Sociological theory
- Statistics: Extreme value theory
- Theatre: Performance theory
- Visual Art: Aesthetics — Art Educational theory — Architecture — Composition — Anatomy — Color theory — Perspective — Visual perception — Geometry — Manifolds
- Other: Obsolete scientific theories

Law (Principle)

A law is a universal principle that describes the fundamental nature of something, the universal properties and the relationships between things, or a description that purports to explain these principles and relationships.

"Laws of Nature"

For example, "physical laws" such as the "law of gravity" (which is in fact more a "force" than a "law"), or "scientific laws" attempt to describe the fundamental nature of the universe itself. Laws of mathematics and logic describe the nature of rational thought and inference (Kant's transcendental idealism, and differently G. Spencer-Brown's work *Laws of Form*, was precisely a determination of the *a priori* laws governing human thought before any interaction whatsoever with experience).

Within most fields of study, and in science in particular, the elevation of some principle of that field to the status of "law" usually takes place after a very long time during which the principle is used and tested and verified; though in some fields of study such laws are simply postulated as a foundation and assumed. Mathematical laws are somewhere in between: they are often arbitrary and unproven in themselves, but they are sometimes judged by how useful they are in making predictions about the real world. However, they ultimately rely on arbitrary axioms.

"Laws" in Social Sciences

Laws of economics are an attempt in modelization of economic behavior. Marxism criticized the belief in eternal "laws of economics", which it considered a product of the dominant ideology. It claimed that in fact, those so-called "laws of economics" were only the historical laws of capitalism, that is of a particular historical social formation. With the advent, in the 20th century, of the application of mathematical, statistical, and experimental techniques to economics, economic theory matured into a corpus of knowledge rooted in the scientific method rather than in philosophical argument.

Miscellaneous

Finally, the term is sometimes applied to less rigorous ideas that may be interesting observations or relationships, practical or ethical guidelines (also called rules of thumb), and even humorous parodies of such laws.

Examples of scientific laws include Boyle's law of gases, conservation laws, Ohm's law, and others. Laws of other fields of study include Occam's razor as a principle of philosophy and Say's law in economics. Examples of observed phenomena often described as laws include the Titius-Bode law of planetary positions, Zipf's law of linguistics, Thomas Malthus's Principle of Population or Malthusian Growth Model, Moore's law of technological growth. Other laws are pragmatic and observational, such as the law of unintended consequences.

Some humorous parodies of such laws include adages such as Murphy's law and its many variants, and Godwin's Law of Internet conversations.

Conceptual Model

A conceptual model is a representation of a system, made of the composition of concepts which are used to help people know, understand, or simulate a subject the model represents. Some models are physical objects; for example, a toy model which may be assembled, and may be made to work like the object it represents.

The term *conceptual model* may be used to refer to models which are formed after a conceptualization or generalization process. Conceptual models are often abstractions of things in the real world whether physical or social. Semantics studies are relevant to various stages of concept formation and use as Semantics is basically about concepts, the meaning that thinking beings give to various elements of their experience.

Models of Concepts and Models that are Conceptual

The term *conceptual model* is normal. It could mean "a model of concept" or it could mean "a model that is conceptual." A distinction can be made between *what models are* and *what models are models of*. With the exception of iconic models, such as a scale model of Winchester Cathedral, most models are concepts. But they are, mostly, intended to be models of real world states of affairs. The value of a model is usually directly proportional to how well it corresponds to a past, present, future, actual or potential state of affairs. A model of a concept is quite different because in order to be a good model it need not have this real world correspondence. In artificial intelligence conceptual models and conceptual graphs are used for building expert systems and knowledge-based systems; here the analysts are concerned to represent expert opinion on what is true not their own ideas on what is true.

Type and Scope of Conceptual Models

Conceptual models (models that are conceptual) range in type from the more concrete, such as the mental image of a familiar physical object, to the formal generality and abstractness of mathematical models which do not appear to the mind as an image. Conceptual models also range in terms of the scope of the subject matter that they are taken to represent. A model may, for instance, represent a single thing (e.g. the *Statue of Liberty*), whole classes of things (e.g. *the electron*), and even very vast domains of subject matter such as *the physical universe*. The variety and scope of conceptual models is due to the variety of purposes had by the people using them.

Overview

Conceptual modeling is the activity of formally describing some aspects of the physical and social world around us for the purposes of understanding and communication."

A conceptual model's primary objective is to convey the fundamental principles and basic functionality of the system which it represents. Also, a conceptual model must be developed in such a way as to provide an easily understood system interpretation for the models users. A conceptual model, when implemented properly, should satisfy four fundamental objectives.

1. Enhance an individual's understanding of the representative system

2. Facilitate efficient conveyance of system details between stakeholders

3. Provide a point of reference for system designers to extract system specifications

4. Document the system for future reference and provide a means for collaboration

The conceptual model plays an important role in the overall system development life cycle. Figure 1 below, depicts the role of the conceptual model in a typical system development scheme. It is clear that if the conceptual model is not fully developed, the execution of fundamental system properties may not be implemented properly, giving way to future problems or system shortfalls. These failures do occur in the industry and have been linked to; lack of user input, incomplete or unclear requirements, and changing requirements. Those weak links in the system design and development process can be traced to improper execution of the fundamental objectives of conceptual modeling. The importance of conceptual modeling is evident when such systemic failures are mitigated by thorough system development and adherence to proven development objectives/techniques.

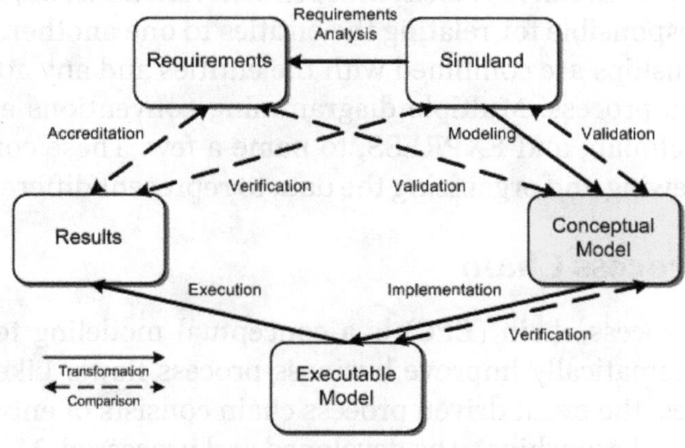

J. Sokolowski, C. Banks, Modeling and Simulation Fundamentals: Theoretical Underpinnings and Practical Domains, Wiley, 2010, pp 333

Techniques

As systems have become increasingly complex, the role of conceptual modeling has dramatically expanded. With that expanded presence, the effectiveness of conceptual modeling at capturing the fundamentals of a system is being realized. Building on that realization, numerous conceptual modeling techniques have been created. These techniques can be applied across multiple disciplines to increase the users understanding of the system to be modeled. A few techniques are briefly described in the following text, however, many more exist or are being developed. Some commonly used conceptual modeling techniques and methods include; Workflow Modeling, Workforce Modeling,

Rapid Application Development, Object Role Modeling, and Unified Modeling Language (UML).

Data Flow Modeling

Data flow modeling (DFM) is a basic conceptual modeling technique that graphically represents elements of a system. DFM is a fairly simple technique, however, like many conceptual modeling techniques, it is possible to construct higher and lower level representative diagrams. The data flow diagram usually does not convey complex system details such as parallel development considerations or timing information, but rather works to bring the major system functions into context. Data flow modeling is a central technique used in systems development that utilizes the Structured Systems Analysis and Design Method (SSADM).

Entity Relationship Modeling

Entity-relationship modeling (ERM) is a conceptual modeling technique used primarily for software system representation. Entity-relationship diagrams, which are a product of executing the ERM technique, are normally used to represent database models and information systems. The main components of the diagram are the entities and relationships. The entities can represent independent functions, objects, or events. The relationships are responsible for relating the entities to one another. To form a system process, the relationships are combined with the entities and any attributes needed to further describe the process. Multiple diagramming conventions exist for this technique; IDEF1X, Bachman, and EXPRESS, to name a few. These conventions are just different ways of viewing and organizing the data to represent different system aspects.

Event-driven Process Chain

The event-driven process chain (EPC) is a conceptual modeling technique which is mainly used to systematically improve business process flows. Like most conceptual modeling techniques, the event driven process chain consists of entities/elements and functions that allow relationships to be developed and processed. More specifically, the EPC is made up of events which define what state a process is in or the rules by which it operates. In order to progress through events, a function/ active event must be executed. Depending on the process flow, the function has the ability to transform event states or link to other event driven process chains. Other elements exist within an EPC, all of which work together to define how and by what rules the system operates. The EPC technique can be applied to business practices such as resource planning, process improvement, and logistics.

Joint Application Development

The Dynamic Systems Development Method (DSDM) uses a specific process called

JEFFF to conceptually model a systems life cycle. JEFFF is intended to focus more on the higher level development planning that precedes a projects initialization. The JAD process calls for a series of workshops in which the participants work to identify, define, and generally map a successful project from conception to completion. This method has been found to not work well for large scale applications, however smaller applications usually report some net gain in efficiency.

Place/Transition Net

Also known as Petri Nets, this conceptual modeling technique allows a system to be constructed with elements that can be described by direct mathematical means. The petri net, because of its nondeterministic execution properties and well defined mathematical theory, is a useful technique for modeling concurrent system behavior, i.e. simultaneous process executions.

State Transition Modeling

State transition modeling makes use of state transition diagrams to describe system behavior. These state transition diagrams use distinct states to define system behavior and changes. Most current modeling tools contain some kind of ability to represent state transition modeling. The use of state transition models can be most easily recognized as logic state diagrams and directed graphs for finite state machines.

Technique Evaluation and Selection

Because the conceptual modeling method can sometimes be purposefully vague to account for a broad area of use, the actual application of concept modeling can become difficult. To alleviate this issue, and shed some light on what to consider when selecting an appropriate conceptual modeling technique, the framework proposed by Gemino and Wand will be discussed in the following text. However, before evaluating the effectiveness of a conceptual modeling technique for a particular application, an important concept must be understood; Comparing conceptual models by way of specifically focusing on their graphical or top level representations is shortsighted. Gemino and Wand make a good point when arguing that the emphasis should be placed on a conceptual modeling language when choosing an appropriate technique. In general, a conceptual model is developed using some form of conceptual modeling technique. That technique will utilize a conceptual modeling language that determines the rules for how the model is arrived at. Understanding the capabilities of the specific language used is inherent to properly evaluating a conceptual modeling technique, as the language reflects the techniques descriptive ability. Also, the conceptual modeling language will directly influence the depth at which the system is capable of being represented, whether it be complex or simple.

Considering affecting factors

Building on some of their earlier work, Gemino and Wand acknowledge some main points to consider when studying the affecting factors: the content that the conceptual model must represent, the method in which the model will be presented, the characteristics of the models users, and the conceptual model languages specific task. The conceptual models content should be considered in order to select a technique that would allow relevant information to be presented. The presentation method for selection purposes would focus on the techniques ability to represent the model at the intended level of depth and detail. The characteristics of the models users or participants is an important aspect to consider. A participant's background and experience should coincide with the conceptual models complexity, else misrepresentation of the system or misunderstanding of key system concepts could lead to problems in that systems realization. The conceptual model language task will further allow an appropriate technique to be chosen. The difference between creating a system conceptual model to convey system functionality and creating a system conceptual model to interpret that functionality could involve to completely different types of conceptual modeling languages.

Considering affected variables

Gemino and Wand go on to expand the affected variable content of their proposed framework by considering the focus of observation and the criterion for comparison. The focus of observation considers whether the conceptual modeling technique will create a "new product", or whether the technique will only bring about a more intimate understanding of the system being modeled. The criterion for comparison would weigh the ability of the conceptual modeling technique to be efficient or effective. A conceptual modeling technique that allows for development of a system model which takes all system variables into account at a high level may make the process of understanding the system functionality more efficient, but the technique lacks the necessary information to explain the internal processes, rendering the model less effective.

When deciding which conceptual technique to use, the recommendations of Gemino and Wand can be applied in order to properly evaluate the scope of the conceptual model in question. Understanding the conceptual models scope will lead to a more informed selection of a technique that properly addresses that particular model. In summary, when deciding between modeling techniques, answering the following questions would allow one to address some important conceptual modeling considerations.

1. What content will the conceptual model represent?
2. How will the conceptual model be presented?
3. Who will be using or participating in the conceptual model?

4. How will the conceptual model describe the system?
5. What is the conceptual models focus of observation?
6. Will the conceptual model be efficient or effective in describing the system?

Another function of the simulation conceptual model is to provide a rational and factual basis for assessment of simulation application appropriateness.

Models in Philosophy and Science

Mental Model

In cognitive psychology and philosophy of mind, a mental model is a representation of something in the mind, but a mental model may also refer to a nonphysical external model of the mind itself.

Metaphysical Models

A metaphysical model is a type of conceptual model which is distinguished from other conceptual models by its proposed scope. A metaphysical model intends to represent reality in the broadest possible way. This is to say that it explains the answers to fundamental questions such as whether matter and mind are one or two substances; or whether or not humans have free will.

Conceptual Model Vs. Semantics Model

Epistemological Models

An epistemological model is a type of conceptual model whose proposed scope is the known and the knowable, and the believed and the believable.

Logical Models

In logic, a model is a type of interpretation under which a particular statement is true. Logical models can be broadly divided into ones which only attempt to represent concepts, such as mathematical models; and ones which attempt to represent physical objects, and factual relationships, among which are scientific models.

Model theory is the study of (classes of) mathematical structures such as groups, fields, graphs, or even universes of set theory, using tools from mathematical logic. A system that gives meaning to the sentences of a formal language is called a model for the language. If a model for a language moreover satisfies a particular sentence or theory (set of sentences), it is called a model of the sentence or theory. Model theory has close ties to algebra and universal algebra.

Mathematical Models

Mathematical models can take many forms, including but not limited to dynamical systems, statistical models, differential equations, or game theoretic models. These and other types of models can overlap, with a given model involving a variety of abstract structures.

A more comprehensive type of mathematical model uses a linguistic version of category theory to model a given situation. Akin to entity-relationship models, custom categories or sketches can be directly translated into database schemas. The difference is that logic is replaced by category theory, which brings powerful theorems to bear on the subject of modeling, especially useful for translating between disparate models (as functors between categories).

Scientific Models

A scientific model is a simplified abstract view of a complex reality. A scientific model represents empirical objects, phenomena, and physical processes in a logical way. Attempts to formalize the principles of the empirical sciences use an interpretation to model reality, in the same way logicians axiomatize the principles of logic. The aim of these attempts is to construct a formal system for which reality is the only interpretation. The world is an interpretation (or model) of these sciences, only insofar as these sciences are true.

Statistical Models

A statistical model is a probability distribution function proposed as generating data. In a parametric model, the probability distribution function has variable parameters, such as the mean and variance in a normal distribution, or the coefficients for the various exponents of the independent variable in linear regression. A nonparametric model has a distribution function without parameters, such as in bootstrapping, and is only loosely confined by assumptions. Model selection is a statistical method for selecting a distribution function within a class of them, e.g., in linear regression where the dependent variable is a polynomial of the independent variable with parametric coefficients, model selection is selecting the highest exponent, and may be done with nonparametric means, such as with cross validation.

In statistics there can be models of mental events as well as models of physical events. For example, a statistical model of customer behavior is a model that is conceptual (because behavior is physical), but a statistical model of customer satisfaction is a model of a concept (because satisfaction is a mental not a physical event).

Social and Political Models

Economic Models

In economics, a model is a theoretical construct that represents economic processes by a set of variables and a set of logical and/or quantitative relationships between them. The economic model is a simplified framework designed to illustrate complex processes, often but not always using mathematical techniques. Frequently, economic models use structural parameters. Structural parameters are underlying parameters in a model or class of models. A model may have various parameters and those parameters may change to create various properties.

Models in Systems Architecture

A system model is the conceptual model that describes and represents the structure, behavior, and more views of a system. A system model can represent multiple views of a system by using two different approaches. The first one is the non-architectural approach and the second one is the architectural approach. The non-architectural approach respectively picks a model for each view. The architectural approach, also known as system architecture, instead of picking many heterogeneous and unrelated models, will use only one integrated architectural model.

Business Process Modelling

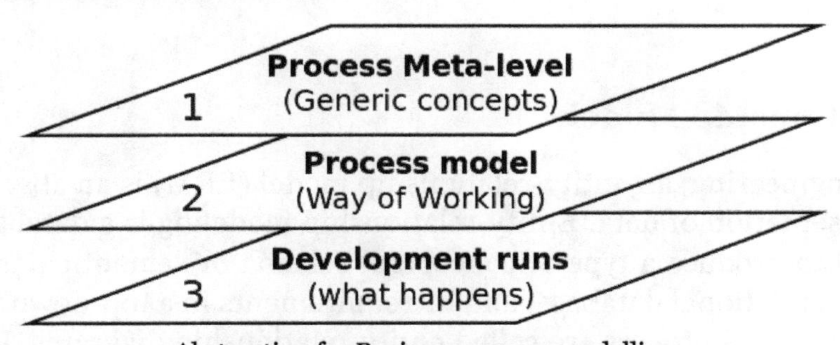

Abstraction for Business process modelling

In business process modelling the enterprise process model is often referred to as the *business process model*. Process models are core concepts in the discipline of process engineering. Process models are:

- Processes of the same nature that are classified together into a model.
- A description of a process at the type level.
- Since the process model is at the type level, a process is an instantiation of it.

The same process model is used repeatedly for the development of many applications and thus, has many instantiations.

One possible use of a process model is to prescribe how things must/should/could be done in contrast to the process itself which is really what happens. A process model is roughly an anticipation of what the process will look like. What the process shall be will be determined during actual system development.

Models in Information System Design

Conceptual Models of Human Activity Systems

Conceptual models of human activity systems are used in Soft systems methodology (SSM) which is a method of systems analysis concerned with the structuring of problems in management. These models are models of concepts; the authors specifically state that they are not intended to represent a state of affairs in the physical world. They are also used in Information Requirements Analysis (IRA) which is a variant of SSM developed for information system design and software engineering.

Logico-linguistic Models

Logico-linguistic modeling is another variant of SSM that uses conceptual models. However, this method combines models of concepts with models of putative real world objects and events. It is a graphical representation of modal logic in which modal operators are used to distinguish statement about concepts from statements about real world objects and events.

Data Models

Entity-relationship Model

In software engineering, an entity-relationship model (ERM) is an abstract and conceptual representation of data. Entity-relationship modeling is a database modeling method, used to produce a type of conceptual schema or semantic data model of a system, often a relational database, and its requirements in a top-down fashion. Diagrams created by this process are called entity-relationship diagrams, ER diagrams, or ERDs.

Entity-relationship models have had wide application in the building of information systems intended to support activities involving objects and events in the real world. In these cases they are models that are conceptual. However, this modeling method can be used to build computer games or a family tree of the Greek Gods, in these cases it would be used to model concepts.

Domain Model

A domain model is a type of conceptual model used to depict the structural elements and their conceptual constraints within a domain of interest (sometimes called the problem domain). A domain model includes the various entities, their attributes and relationships, plus the constraints governing the conceptual integrity of the structural model elements comprising that problem domain. A domain model may also include a number of conceptual views, where each view is pertinent to a particular subject area of the domain or to a particular subset of the domain model which is of interest to a stakeholder of the domain model.

Like entity-relationship models, domain models can be used to model concepts or to model real world objects and events.

Generalization

A generalization (or generalisation) is a concept in the inductive sense of that word, or an extension of a concept to less-specific English or mathematical criteria. Generalizations posit the existence of a domain or set of elements, as well as one or more common characteristics shared by those elements (thus creating a conceptual model). As such, they are the essential basis of all valid deductive inferences. The process of verification is necessary to determine whether a generalization holds true for any given situation.

The concept of generalization has broad application in many related disciplines, sometimes having a specialized context or meaning.

Of any two related concepts, such as *A* and *B*, *A* is a "generalization" of *B*, and *B* is a special case of *A*, if and only if

- every instance of concept *B* is also an instance of concept *A*; and
- there are instances of concept *A* which are not instances of concept *B*.

For instance, *animal* is a generalization of *bird* because every bird is an animal, and there are animals which are not birds (dogs, for instance)

Hypernym and Hyponym

The relation of *generalization* to *specialization* (or *particularization*) is reflected in the contrasting words hypernym and hyponym. A hypernym as a generic stands for a class or group of equally ranked items—for example, *tree* stands for equally ranked items such as *peach* and *oak*, and *ship* stands for equally ranked items such as *cruiser* and *steamer*. In contrast, a hyponym is one of the items included in the generic, such as *peach* and *oak* which are included in *tree*, and *cruiser* and *steamer* which are included in *ship*. A hypernym is superordinate to a hyponym, and a hyponym is subordinate to a hypernym.

Examples

Biological Generalization

An animal is a generalization of a mammal, a bird, a fish, an amphibian and a reptile.

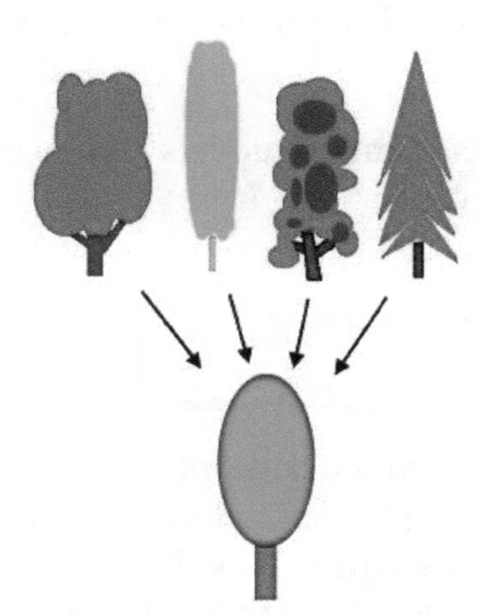

When the mind makes a generalization, it extracts the essence of a concept based on its analysis of similarities from many discrete objects. The resulting simplification enables higher-level thinking.

Cartographic Generalization of Geo-spatial Data

Generalization has a long history in cartography as an art of creating maps for different scale and purpose. Cartographic generalization is the process of selecting and representing information of a map in a way that adapts to the scale of the display medium of the map. In this way, every map has, to some extent, been generalized to match the criteria of display. This includes small cartographic scale maps, which cannot convey every detail of the real world. Cartographers must decide and then adjust the content within their maps to create a suitable and useful map that conveys geospatial information within their representation of the world.

Generalization is meant to be context-specific. That is to say, correctly generalized maps are those that emphasize the most important map elements while still representing the world in the most faithful and recognizable way. The level of detail and importance in what is remaining on the map must outweigh the insignificance of items that were generalized, as to preserve the distinguishing characteristics of what makes the map useful and important.

Geometric Generalizations

A polygon is a generalization of a 3-sided triangle, a 4-sided quadrilateral, and so on to *n* sides.

A hypercube is a generalization of a 2-dimensional square, a 3-dimensional cube, and so on to *n* dimensions.

A quadric, such as a hypersphere, ellipsoid, paraboloid, or hyperboloid, is a generalization of a conic section to higher dimensions.

References

- Schick, Theodore; Vaughn, Lewis (2002). How to think about weird things: critical thinking for a New Age. Boston: McGraw-Hill Higher Education. ISBN 0-7674-2048-9.
- Hilborn, Ray; Mangel, Marc (1997). The ecological detective: confronting models with data. Princeton University Press. p. 24. ISBN 978-0-691-03497-3. Retrieved 22 August 2011.
- Patricia M. Shields (1998). "Pragmatism As a Philosophy of Science: A Tool For Public Administration". In Jay D. White. Research in Public Administration. 4. pp. 195–225 [211]. ISBN 1-55938-888-9.
- J. Sokolowski, C. Banks, Modeling and Simulation Fundamentals: Theoretical Underpinnings and Practical Domains, Wiley (2010). Amazon.com. ISBN 0470486740.

6

Essential Aspects of Research Design

Observational study is the study of a sample to a population. In this study the subject is not under the control of the researcher mainly because of ethical concerns. The other essential aspects that have been explained are experiments, field experiments, literature review and systematic review. The major categories of research design are dealt with great detail in the chapter.

Observational Study

In fields such as epidemiology, social sciences, psychology and statistics, an observational study draws inferences from a sample to a population where the independent variable is not under the control of the researcher because of ethical concerns or logistical constraints. One common observational study is about the possible effect of a treatment on subjects, where the assignment of subjects into a treated group versus a control group is outside the control of the investigator. This is in contrast with experiments, such as randomized controlled trials, where each subject is randomly assigned to a treated group or a control group.

Weaknesses

The independent variable may be beyond the control of the investigator for a variety of reasons:

- A randomized experiment would violate ethical standards. Suppose one wanted to investigate the abortion – breast cancer hypothesis, which postulates a causal link between induced abortion and the incidence of breast cancer. In a hypothetical controlled experiment, one would start with a large subject pool of pregnant women and divide them randomly into a treatment group (receiving induced abortions) and a control group (not receiving abortions), and then conduct regular cancer screenings for women from both groups. Needless to say, such an experiment would run counter to common ethical principles. (It would also suffer from various confounds and sources of bias, e.g. it would be impossible to conduct it as a blind experiment.) The published studies investigating the abortion–breast cancer hypothesis generally start with a group of women who already have received abortions. Membership in this "treated" group is not controlled by the investigator: the group is formed after the "treatment" has been assigned.

- The investigator may simply lack the requisite influence. Suppose a scientist wants to study the public health effects of a community-wide ban on smoking in public indoor areas. In a controlled experiment, the investigator would randomly pick a set of communities to be in the treatment group. However, it is typically up to each community and/or its legislature to enact a smoking ban. The investigator can be expected to lack the political power to cause precisely those communities in the randomly selected treatment group to pass a smoking ban. In an observational study, the investigator would typically start with a treatment group consisting of those communities where a smoking ban is already in effect.
- A randomized experiment may be impractical. Suppose a researcher wants to study the suspected link between a certain medication and a very rare group of symptoms arising as a side effect. Setting aside any ethical considerations, a randomized experiment would be impractical because of the rarity of the effect. There may not be a subject pool large enough for the symptoms to be observed in at least one treated subject. An observational study would typically start with a group of symptomatic subjects and work backwards to find those who were given the medication and later developed the symptoms. Thus a subset of the treated group was determined based on the presence of symptoms, instead of by random assignment.

Types of Observational Studies

- Case-control study: study originally developed in epidemiology, in which two existing groups differing in outcome are identified and compared on the basis of some supposed causal attribute.
- Cross-sectional study: involves data collection from a population, or a representative subset, at one specific point in time.
- Longitudinal study: correlational research study that involves repeated observations of the same variables over long periods of time.
- Cohort study or Panel study: a particular form of longitudinal study where a group of patients is closely monitored over a span of time.
- Ecological study: an observational study in which at least one variable is measured at the group level.

Degree of Usefulness and Reliability

Although observational studies cannot be used as reliable sources to make statements of fact about the "safety, efficacy, or effectiveness" of a practice, they can still be of use for some other things:

> "[T]hey can: 1) provide information on "real world" use and practice; 2) detect signals about the benefits and risks of...[the] use [of practices] in the general population; 3) help formulate hypotheses to be tested in subsequent experiments; 4) provide part of the community-level data needed to design more informative pragmatic clinical trials; and 5) inform clinical practice."

Bias and Compensating Methods

In all of those cases, if a randomized experiment cannot be carried out, the alternative line of investigation suffers from the problem that the decision of which subjects receive the treatment is not entirely random and thus is a potential source of bias. A major challenge in conducting observational studies is to draw inferences that are acceptably free from influences by overt biases, as well as to assess the influence of potential hidden biases.

An observer of an uncontrolled experiment (or process) records potential factors and the data output: the goal is to determine the effects of the factors. Sometimes the recorded factors may not be directly causing the differences in the output. There may be more important factors which were not recorded but are, in fact, causal. Also, recorded or unrecorded factors may be correlated which may yield incorrect conclusions. Finally, as the number of recorded factors increases, the likelihood increases that at least one of the recorded factors will be highly correlated with the data output simply by chance.

In lieu of experimental control, multivariate statistical techniques allow the approximation of experimental control with statistical control, which accounts for the influences of observed factors that might influence a cause-and-effect relationship. In healthcare and the social sciences, investigators may use matching to compare units that nonrandomly received the treatment and control. One common approach is to use propensity score matching in order to reduce confounding.

A report from the Cochrane Collaboration in 2014 came to the conclusion that observational studies are very similar in results reported by similarly conducted randomized controlled trials. In other words, it reported little evidence for significant effect estimate differences between observational studies and randomized controlled trials, regardless of specific observational study design, heterogeneity, or inclusion of studies of pharmacological interventions. It therefore recommended that factors other than study design *per se* need to be considered when exploring reasons for a lack of agreement between results of randomized controlled trials and observational studies.

In 2007, several prominent medical researchers issued the *Strengthening the reporting of observational studies in epidemiology* (STROBE) statement, in which they called for observational studies to conform to 22 criteria that would make their conclusions easier to understand and generalise.

Case-control Study

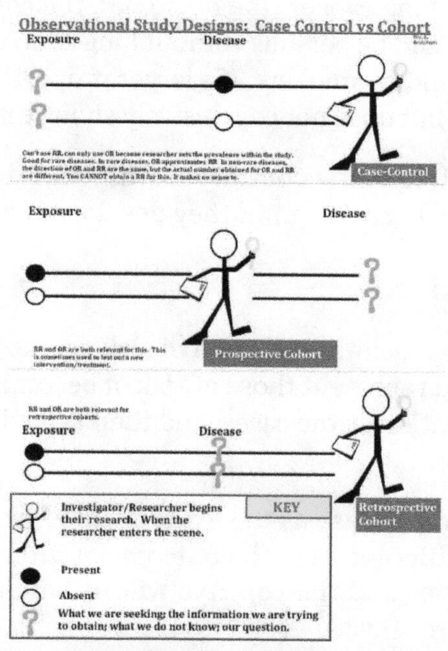

Case-Control Study vs. Cohort on a Timeline. "OR" stands for "odds ratio" and "RR" stands for "relative risk".

A case-control study is a type of observational study in which two existing groups differing in outcome are identified and compared on the basis of some supposed causal attribute. Case-control studies are often used to identify factors that may contribute to a medical condition by comparing subjects who have that condition/disease (the "cases") with patients who do not have the condition/disease but are otherwise similar (the "controls"). They require fewer resources but provide less evidence for causal inference than a randomized controlled trial.

Definition

The case-control is a type of epidemiological observational study. An observational study is a study in which subjects are not randomized to the exposed or unexposed groups, rather the subjects are *observed* in order to determine both their exposure and their outcome status and the exposure status is thus not determined by the researcher.

Porta's *Dictionary of Epidemiology* defines the case-control study as: an observational epidemiological study of persons with the disease (or another outcome variable) of interest and a suitable control group of persons without the disease (comparison group, reference group). The potential relationship of a suspected risk factor or an attribute to the disease is examined by comparing the diseased and nondiseased subjects with regard to how frequently the factor or attribute is present (or, if quantitative, the levels of the attribute) in each of the groups (diseased and nondiseased)."

For example, in a study trying to show that people who smoke (the *attribute*) are more likely to be diagnosed with lung cancer (the *outcome*), the *cases* would be persons with lung cancer, the *controls* would be persons without lung cancer (not necessarily healthy), and some of each group would be smokers. If a larger proportion of the cases smoke than the controls, that suggests, but does not conclusively show, that the hypothesis is valid.

The case-control study is frequently contrasted with cohort studies, wherein exposed and unexposed subjects are observed until they develop an outcome of interest.

Control Group Selection

Controls need not be in good health; inclusion of sick people is sometimes appropriate, as the control group should represent those at risk of becoming a case. Controls should come from the same population as the cases, and their selection should be independent of the exposures of interest.

Controls can carry the same disease as the experimental group, but of another grade/severity, therefore being different from the outcome of interest. However, because the difference between the cases and the controls will be smaller, this results in a lower power to detect an exposure effect.

As with any epidemiological study, greater numbers in the study will increase the power of the study. Numbers of cases and controls do not have to be equal. In many situations, it is much easier to recruit controls than to find cases. Increasing the number of controls above the number of cases, up to a ratio of about 4 to 1, may be a cost-effective way to improve the study.

Strengths and Weaknesses

Case-control studies are a relatively inexpensive and frequently used type of epidemiological study that can be carried out by small teams or individual researchers in single facilities in a way that more structured experimental studies often cannot be. They have pointed the way to a number of important discoveries and advances. The case-control study design is often used in the study of rare diseases or as a preliminary study where little is known about the association between the risk factor and disease of interest.

Compared to prospective cohort studies they tend to be less costly and shorter in duration. In several situations they have greater statistical power than cohort studies, which must often wait for a 'sufficient' number of disease events to accrue.

Case-control studies are observational in nature and thus do not provide the same level of evidence as randomized controlled trials. The results may be confounded by other factors, to the extent of giving the opposite answer to better studies. A meta-analysis of what were considered 30 high-quality studies concluded that use of a product halved a risk, when in fact the risk was, if anything, increased. It may also be more difficult to establish the timeline of exposure to disease outcome in the setting of a case-control study than within a prospective cohort study design where the exposure is ascertained prior to following the subjects over time in order to ascertain their outcome status. The most important drawback in case-control studies relates to the difficulty of obtaining reliable information about an individual's exposure status over time. Case-control studies are therefore placed low in the hierarchy of evidence.

Examples

One of the most significant triumphs of the case-control study was the demonstration of the link between tobacco smoking and lung cancer, by Richard Doll and Bradford Hill. They showed a statistically significant association in a large case-control study. Opponents argued for many years that this type of study cannot prove causation, but the eventual results of cohort studies confirmed the causal link which the case-control studies suggested, and it is now accepted that tobacco smoking is the cause of about 87% of all lung cancer mortality in the US.

Analysis

Case-control studies were initially analyzed by testing whether or not there were significant differences between the proportion of exposed subjects among cases and controls. Subsequently Cornfield pointed out that, when the disease outcome of interest is rare, the odds ratio of exposure can be used to estimate the relative risks. It was later shown by Miettinen in 1976 that this assumption is not necessary and that the odds ratio of exposure can be used to directly estimate the incidence rate ratio of exposure without the need for the rare disease assumption.

Experiment

Even very young children perform rudimentary experiments to learn about the world and how things work.

An experiment is a procedure carried out to support, refute, or validate a hypothesis. Experiments provide insight into cause-and-effect by demonstrating what outcome occurs when a particular factor is manipulated. Experiments vary greatly in goal and scale, but always rely on repeatable procedure and logical analysis of the results. There also exists natural experimental studies.

A child may carry out basic experiments to understand gravity, while teams of scientists may take years of systematic investigation to advance their understanding of a phenomenon. Experiments and other types of hands-on activities are very important to student learning in the science classroom. Experiments can raise test scores and help a student become more engaged and interested in the material they are learning, especially when used over time. Experiments can vary from personal and informal natural comparisons (e.g. tasting a range of chocolates to find a favorite), to highly controlled (e.g. tests requiring complex apparatus overseen by many scientists that hope to discover information about subatomic particles). Uses of experiments vary considerably between the natural and human sciences.

Experiments typically include controls, which are designed to minimize the effects of variables other than the single independent variable. This increases the reliability of the results, often through a comparison between control measurements and the other measurements. Scientific controls are a part of the scientific method. Ideally, all variables in an experiment are controlled (accounted for by the control measurements) and none are uncontrolled. In such an experiment, if all controls work as expected, it is possible to conclude that the experiment works as intended, and that results are due to the effect of the tested variable.

Overview

In the scientific method, an experiment is an empirical procedure that arbitrates between competing models or hypotheses. Researchers also use experimentation to test existing theories or new hypotheses to support or disprove them.

An experiment usually tests a hypothesis, which is an expectation about how a particular process or phenomenon works. However, an experiment may also aim to answer a "what-if" question, without a specific expectation about what the experiment reveals, or to confirm prior results. If an experiment is carefully conducted, the results usually either support or disprove the hypothesis. According to some philosophies of science, an experiment can never "prove" a hypothesis, it can only add support. On the other hand, an experiment that provides a counterexample can disprove a theory or hypothesis. An experiment must also control the possible confounding factors—any factors that would mar the accuracy or repeatability of the experiment or the ability to interpret the results. Confounding is commonly eliminated through scientific controls and/or, in randomized experiments, through random assignment.

In engineering and the physical sciences, experiments are a primary component of the scientific method. They are used to test theories and hypotheses about how physical processes work under particular conditions (e.g., whether a particular engineering process can produce a desired chemical compound). Typically, experiments in these fields focus on replication of identical procedures in hopes of producing identical results in each replication. Random assignment is uncommon.

In medicine and the social sciences, the prevalence of experimental research varies widely across disciplines. When used, however, experiments typically follow the form of the clinical trial, where experimental units (usually individual human beings) are randomly assigned to a treatment or control condition where one or more outcomes are assessed. In contrast to norms in the physical sciences, the focus is typically on the average treatment effect (the difference in outcomes between the treatment and control groups) or another test statistic produced by the experiment. A single study typically does not involve replications of the experiment, but separate studies may be aggregated through systematic review and meta-analysis.

There are various differences in experimental practice in each of the branches of science. For example, agricultural research frequently uses randomized experiments (e.g., to test the comparative effectiveness of different fertilizers), while experimental economics often involves experimental tests of theorized human behaviors without relying on random assignment of individuals to treatment and control conditions.

History

Francis Bacon (1561–1626), an English philosopher and scientist active in the 17th century, became an early and influential supporter of experimental science. He disagreed with the method of answering scientific questions by deduction and described it as follows: “Having first determined the question according to his will, man then resorts to experience, and bending her to conformity with his placets, leads her about like a captive in a procession.” Bacon wanted a method that relied on repeatable observations, or experiments. Notably, he first ordered the scientific method as we understand it today.

There remains simple experience; which, if taken as it comes, is called accident, if sought for, experiment. The true method of experience first lights the candle [hypothesis], and then by means of the candle shows the way [arranges and delimits the experiment]; commencing as it does with experience duly ordered and digested, not bungling or erratic, and from it deducing axioms [theories], and from established axioms again new experiments.

In the centuries that followed, people who applied the scientific method in different areas made important advances and discoveries. For example, Galileo Galilei (1564-

1642) accurately measured time and experimented to make accurate measurements and conclusions about the speed of a falling body. Antoine Lavoisier (1743-1794), a French chemist, used experiment to describe new areas, such as combustion and biochemistry and to develop the theory of conservation of mass (matter). Louis Pasteur (1822-1895) used the scientific method to disprove the prevailing theory of spontaneous generation and to develop the germ theory of disease. Because of the importance of controlling potentially confounding variables, the use of well-designed laboratory experiments is preferred when possible.

A considerable amount of progress on the design and analysis of experiments occurred in the early 20th century, with contributions from statisticians such as Ronald Fisher (1890-1962), Jerzy Neyman (1894-1981), Oscar Kempthorne (1919-2000), Gertrude Mary Cox (1900-1978), and William Gemmell Cochran (1909-1980), among others.

Types of Experiment

Experiments might be categorized according to a number of dimensions, depending upon professional norms and standards in different fields of study. In some disciplines (e.g., psychology or political science), a 'true experiment' is a method of social research in which there are two kinds of variables. The independent variable is manipulated by the experimenter, and the dependent variable is measured. The signifying characteristic of a true experiment is that it randomly allocates the subjects to neutralize experimenter bias, and ensures, over a large number of iterations of the experiment, that it controls for all confounding factors.

Controlled Experiments

A controlled experiment often compares the results obtained from experimental samples against *control* samples, which are practically identical to the experimental sample except for the one aspect whose effect is being tested (the independent variable). A good example would be a drug trial. The sample or group receiving the drug would be the experimental group (treatment group); and the one receiving the placebo or regular treatment would be the control one. In many laboratory experiments it is good practice to have several replicate samples for the test being performed and have both a positive control and a negative control. The results from replicate samples can often be averaged, or if one of the replicates is obviously inconsistent with the results from the other samples, it can be discarded as being the result of an experimental error (some step of the test procedure may have been mistakenly omitted for that sample). Most often, tests are done in duplicate or triplicate. A positive control is a procedure similar to the actual experimental test but is known from previous experience to give a positive result. A negative control is known to give a negative result. The positive control confirms that the basic conditions of the experiment were able to produce a positive result, even if none of the actual experimental samples produce a positive result. The negative control demonstrates the base-line result obtained when a test does not produce a measurable positive result. Most often the value of the negative control is treated as a "background" value to subtract from the test sample results. Sometimes the positive control takes the quadrant of a standard curve.

An example that is often used in teaching laboratories is a controlled protein assay. Students might be given a fluid sample containing an unknown (to the student) amount of protein. It is their job to correctly perform a controlled experiment in which they determine the concentration of protein in fluid sample (usually called the "unknown sample"). The teaching lab would be equipped with a protein standard solution with a known protein concentration. Students could make several positive control samples containing various dilutions of the protein standard. Negative control samples would contain all of the reagents for the protein assay but no protein. In this example, all samples are performed in duplicate. The assay is a colorimetric assay in which a spectrophotometer can measure the amount of protein in samples by detecting a colored complex formed by the interaction of protein molecules and molecules of an added dye. In the illustration, the results for the diluted test samples can be compared to the results of the standard curve (the blue line in the illustration) to estimate the amount of protein in the unknown sample.

Controlled experiments can be performed when it is difficult to exactly control all the conditions in an experiment. In this case, the experiment begins by creating two or more sample groups that are *probabilistically equivalent*, which means that measurements of traits should be similar among the groups and that the groups should respond in the same manner if given the same treatment. This equivalency is determined by statistical methods that take into account the amount of variation between individuals and the number of individuals in each group. In fields such as microbiology and chemistry, where there is very little variation between individuals and the group size is easily in the millions, these statistical methods are often bypassed and simply splitting a solution into equal parts is assumed to produce identical sample groups.

Once equivalent groups have been formed, the experimenter tries to treat them identically except for the one *variable* that he or she wishes to isolate. Human experimentation requires special safeguards against outside variables such as the *placebo effect*. Such experiments are generally *double blind*, meaning that neither the volunteer nor the researcher knows which individuals are in the control group or the experimental group until after all of the data have been collected. This ensures that any effects on the volunteer are due to the treatment itself and are not a response to the knowledge that he is being treated.

In human experiments, researchers may give a subject (person) a stimulus that the subject responds to. The goal of the experiment is to measure the response to the stimulus by a test method.

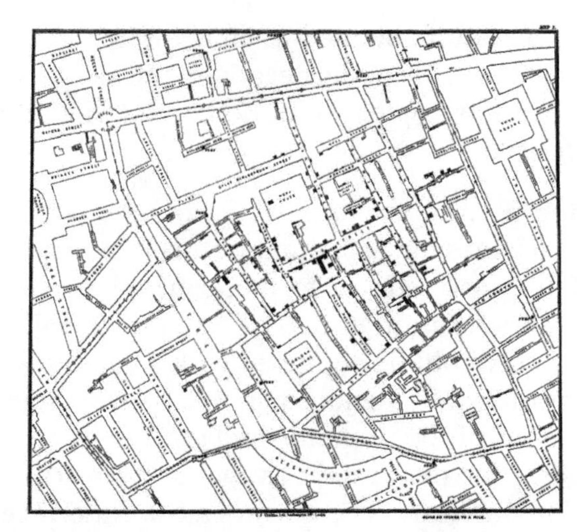

Original map by John Snow showing the clusters of cholera cases in the London epidemic of 1854

In the design of experiments, two or more "treatments" are applied to estimate the difference between the mean responses for the treatments. For example, an experiment on baking bread could estimate the difference in the responses associated with quantitative variables, such as the ratio of water to flour, and with qualitative variables, such as strains of yeast. Experimentation is the step in the scientific method that helps people decide between two or more competing explanations – or hypotheses. These hypotheses suggest reasons to explain a phenomenon, or predict the results of an action. An example might be the hypothesis that "if I release this ball, it will fall to the floor": this suggestion can then be tested by carrying out the experiment of letting go of the ball, and observing the results. Formally, a hypothesis is compared against its opposite or null hypothesis ("if I release this ball, it will not fall to the floor"). The null hypothesis is that there is no explanation or predictive power of the phenomenon through the reasoning that is being investigated. Once hypotheses are defined, an experiment can be carried out and the results analysed to confirm, refute, or define the accuracy of the hypotheses.

Natural Experiments

The term "experiment" usually implies a controlled experiment, but sometimes controlled experiments are prohibitively difficult or impossible. In this case researchers resort to *natural experiments* or *quasi-experiments*. Natural experiments rely solely on observations of the variables of the system under study, rather than manipulation of just one or a few variables as occurs in controlled experiments. To the degree possible, they attempt to collect data for the system in such a way that contribution from all variables can be determined, and where the effects of variation in certain variables remain approximately constant so that the effects of other variables can be discerned. The degree to which this is possible depends on the observed correlation between explanatory variables in the observed data. When these variables are *not* well correlated, natural experiments can approach the power of controlled experiments. Usually, however, there is some correlation between these variables, which reduces the reliability of natural experiments relative to what could be concluded if a controlled experiment were performed. Also, because natural experiments usually take place in uncontrolled environments, variables from undetected sources are neither measured nor held constant, and these may produce illusory correlations in variables under study.

Much research in several science disciplines, including economics, political science, geology, paleontology, ecology, meteorology, and astronomy, relies on quasi-experiments. For example, in astronomy it is clearly impossible, when testing the hypothesis "suns are collapsed clouds of hydrogen," to start out with a giant cloud of hydrogen, and then perform the experiment of waiting a few billion years for it to form a sun. However, by observing various clouds of hydrogen in various states of collapse, and other implications of the hypothesis (for example, the presence of various spectral emissions from the light of stars), we can collect data we require to support the hypothesis. An early example of this type of experiment was the first verification in the 17th century that light does not travel from place to place instantaneously, but instead has a measurable speed. Observation of the appearance of the moons of Jupiter were slightly delayed when Jupiter was farther from Earth, as opposed to when Jupiter was closer to Earth; and this phenomenon was used to demonstrate that the difference in the time of appearance of the moons was consistent with a measurable speed.

Field Experiments

Field experiments are so named to distinguish them from laboratory experiments, which enforce scientific control by testing a hypothesis in the artificial and highly controlled setting of a laboratory. Often used in the social sciences, and especially in economic analyses of education and health interventions, field experiments have the advantage that outcomes are observed in a natural setting rather than in a contrived laboratory environment. For this reason, field experiments are sometimes seen as having higher external validity than laboratory experiments. However, like natural experiments, field experiments suffer from the possibility of contamination: experimental conditions can be controlled with more precision and certainty in the lab. Yet some phenomena (e.g., voter turnout in an election) cannot be easily studied in a laboratory.

Contrast with Observational Study

Experiment

An observational study is used when it is impractical, unethical, cost-prohibitive (or otherwise inefficient) to fit a physical or social system into a laboratory setting, to completely control confounding factors, or to apply random assignment. It can also be used when confounding factors are either limited or known well enough to analyze the data in light of them (though this may be rare when social phenomena are under examination). For an observational science to be valid, the experimenter must know and account for confounding factors. In these situations, observational studies have value because they often suggest hypotheses that can be tested with randomized experiments or by collecting fresh data.

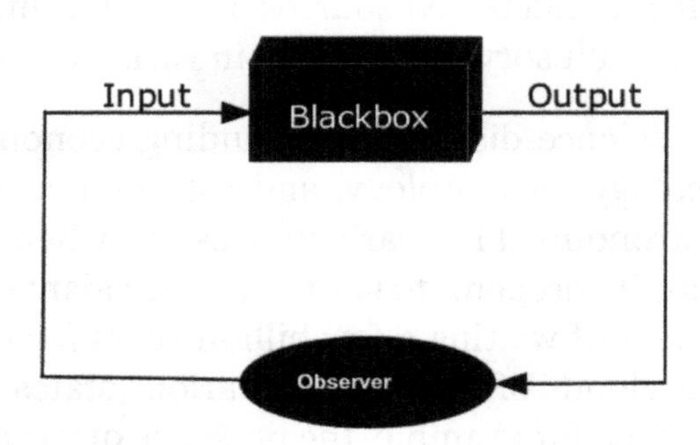

The black box model for observation (input and output are *observables*). When there are a feedback with some observer's control, as illustred, the observation is also an experiment.

Fundamentally, however, observational studies are not experiments. By definition, observational studies lack the manipulation required for Baconian experiments. In addition, observational studies (e.g., in biological or social systems) often involve variables that are difficult to quantify or control. Observational studies are limited because they lack the statistical properties of randomized experiments. In a randomized experiment, the method of randomization specified in the experimental protocol guides the statistical analysis, which is usually specified also by the experimental protocol. Without a statistical model that reflects an objective randomization, the statistical analysis relies on a subjective model. Inferences from subjective models are unreliable in theory and practice. In fact, there are several cases where carefully conducted observational studies consistently give wrong results, that is, where the results of the observational studies are inconsistent and also differ from the results of experiments. For example, epidemiological studies of colon cancer consistently show beneficial correlations with broccoli consumption, while experiments find no benefit.

A particular problem with observational studies involving human subjects is the great difficulty attaining fair comparisons between treatments (or exposures), because such studies are prone to selection bias, and groups receiving different treatments (exposures) may differ greatly according to their covariates (age, height, weight, medications, exercise, nutritional status, ethnicity, family medical history, etc.). In contrast, randomization implies that for each covariate, the mean for each group is expected to be the same. For any randomized trial, some variation from the mean is expected, of course, but the randomization ensures that the experimental groups have mean values that are close, due to the central limit theorem and Markov's inequality. With inadequate randomization or low sample size, the systematic variation in covariates between the treatment groups (or exposure groups) makes it difficult to separate the effect of the treatment (exposure) from the effects of the other covariates, most of which have not been measured. The mathematical models used to analyze such data must consider each differing covariate (if measured), and results are not meaningful if a covariate is neither randomized nor included in the model.

To avoid conditions that render an experiment far less useful, physicians conducting medical trials—say for U.S. Food and Drug Administration approval—quantify and randomize the covariates that can be identified. Researchers attempt to reduce the biases of observational studies with complicated statistical methods such as propensity score matching methods, which require large populations of subjects and extensive information on covariates. Outcomes are also quantified when possible (bone density, the amount of some cell or substance in the blood, physical strength or endurance, etc.) and not based on a subject's or a professional observer's opinion. In this way, the design of an observational study can render the results more objective and therefore, more convincing.

Ethics

By placing the distribution of the independent variable(s) under the control of the researcher, an experiment - particularly when it involves human subjects - introduces potential ethical considerations, such as balancing benefit and harm, fairly distributing interventions (e.g., treatments for a disease), and informed consent. For example, in psychology or health care, it is unethical to provide a substandard treatment to patients. Therefore, ethical review boards are supposed to stop clinical trials and other experiments unless a new treatment is believed to offer benefits as good as current best practice. It is also generally unethical (and often illegal) to conduct randomized experiments on the effects of substandard or harmful treatments, such as the effects of ingesting arsenic on human health. To understand the effects of such exposures, scientists sometimes use observational studies to understand the effects of those factors.

Even when experimental research does not directly involve human subjects, it may still present ethical concerns. For example, the nuclear bomb experiments conducted by the Manhattan Project implied the use of nuclear reactions to harm human beings even though the experiments did not directly involve any human subjects.

Field Experiment

A field experiment applies the scientific method to experimentally examine an intervention in the real world (or as many experimentalists like to say, naturally occurring environments) rather than in the laboratory. Field experiments, like lab experiments, generally randomize subjects (or other sampling units) into treatment and control groups and compare outcomes between these groups. Field experiments are so named in order to draw a contrast with laboratory experiments, which enforce scientific control by testing a hypothesis in the artificial and highly controlled setting of a laboratory. Often used in the social sciences, and especially in economic analyses of education and health interventions, field experiments have the advantage that outcomes are observed in a natural setting rather than in a contrived laboratory environment. For this reason, field experiments are sometimes seen as having higher external validity than laboratory experiments. However, like natural experiments, field experiments suffer from the possibility of contamination: experimental conditions can be controlled with more precision and certainty in the lab. Yet some phenomena (e.g., voter turnout in an election) cannot be easily studied in a laboratory.

Examples include:

- Clinical trials of pharmaceuticals are one example of field experiments.
- Economists have used field experiments to analyze discrimination, health care programs, charitable fundraising, education, information aggregation in markets, and microfinance programs.
- Engineers often conduct field tests of prototype products to validate earlier laboratory tests and to obtain broader feedback.

History

The use of experiments in the lab and the field have a long history in the physical, natural, and life sciences. Geology has a long history of field experiments, since the time of Avicenna, while field experiments in anthropology date back to Biruni's study of India. Social psychology also has a history of field experiments, including work by pioneering figures Philip Zimbardo, Kurt Lewin and Stanley Milgram. In economics, Peter Bohm, University of Stockholm, was one of the first economists to take the tools of experimental economic methods and attempt to try them with field subjects. In the area of development economics, the pioneer work of Hans Binswanger in the late 1970s conducting experiments in India on risk behavior should also be noted. The use of field experiments in economics has grown recently with the work of John A. List, Jeff Carpenter, Juan-Camilo Cardenas, Abigail Barr, Catherine Eckel, Michael Kremer, Paul Gertler, Glenn Harrison, Colin Camerer, Bradley Ruffle, Abhijit Banerjee Esther Duflo, Dean Karlan, Edward "Ted" Miguel, Sendhil Mullainathan, David H. Reiley, among others. Pa124

Field Experiments in International Development Research

Development economists have used field experiments to measure the effectiveness of poverty and health programs in developing countries. Organizations such as the Abdul Latif Jameel Poverty Action Lab (J-PAL) at the Massachusetts Institute of Technology, the Center of Evaluation for Global Action at the University of California, and Innovations for Poverty Action (IPA) in particular have received attention for their use of randomized field experiments to evaluate development programs. The aim of field experiments used in development research is to find causal relationships between policy interventions and development outcomes. Field experiments are seen by some academics as a rigorous way of testing general theories about economic and political behavior and most recently, field experiments have been used by political scientists to study political behavior, institutional dynamics, and conflict in the developing world.

In a randomized field experiment on an international development intervention, researchers would separate participants into two or more groups: a treatment group (or groups) and a control group. Members of the treatment group(s) then receive a particular development intervention being evaluated while the control group does not. (Often the control group receives the intervention later in the roll out of the study.) Field experiments have gained popularity in the field because they allow researchers to guard against selection bias, a problem present in many current studies of development interventions. Selection bias refers to the fact that, in non-experimental settings, the group receiving a development intervention is likely different from a group that is not receiving the intervention. This may occur because of characteristics that make some people more likely to opt into a program, or because of program targeting. Some academics dispute the claim that findings from field experiments are sufficient for establishing and testing theories about behavior. In particular, a hotly contested issue with regards to field experiments is their external validity. Given that field experiments necessarily take place in a specific geographic and political setting, the extent to which findings can be extrapolated to formulate a general theory regarding economic behavior is a concern.

Caveats

- Fairness of randomization (e.g. in 'negative income tax' experiments communities may lobby for their community to get a cash transfer so the assignment is not purely random)
- Contamination of the randomization
- General equilibrium and "scaling-up"
- Difficulty of replicability (field experiments often require special access or permission, or technical detail—e.g., the instructions for precisely how to replicate a field experiment are rarely if ever available in economics)
- Limits on ability to obtain informed consent of participants
- Field testing is always less controlled than laboratory testing. This increases the variability of the types and magnitudes of stress in field testing. The resulting data, therefore, are more varied: larger standard deviation, less precision and accuracy, etc. This leads to the use of larger sample sizes for field testing.

Literature Review

A literature review is a text of a scholarly paper, which includes the current knowledge including substantive findings, as well as theoretical and methodological contributions to a particular topic. Literature reviews are secondary sources, and do not report new or original experimental work. Literature reviews are a basis for research in nearly every academic field. A narrow-scope literature review may be included as part of a peer-reviewed journal article presenting new research, serving to situate the current study within the body of the relevant literature and to provide context for the reader. In such a case, the review usually precedes the methodology and results sections of the work.

Producing a literature review may also be part of graduate and post-graduate student work, including in the preparation of a thesis, dissertation, or a journal article. Literature reviews are also common in a research proposal or prospectus (the document that is approved before a student formally begins a dissertation or thesis).

Types

The main types of literature reviews are: *evaluative*, *exploratory*, and *instrumental*.

A fourth type, the *systematic* review, is often classified separately, but is essentially a literature review focused on a research question, trying to identify, appraise, select and synthesize all high-quality research evidence and arguments relevant to that question. A *meta-analysis* is typically a systematic review using statistical methods to effectively combine the data used on all selected studies to produce a more reliable result.

Process and Product

Shields and Rangarajan (2013) distinguish between the *process* of reviewing the literature and a finished work or *product* known as a literature review. The *process* of reviewing the literature is often ongoing and informs many aspects of the empirical research project. All of the latest literature should inform a research project. Scholars need to be scanning the literature long after a formal literature review *product* appears to be completed.

A careful literature review is usually 15 to 30 pages and could be longer. The process of reviewing the literature requires different kinds of activities and ways of thinking. Shields and Rangarajan (2013) and Granello (2001) link the activities of doing a literature review with Benjamin Bloom's revised taxonomy of the cognitive domain (ways of thinking: remembering, understanding, applying, analyzing, evaluating, and creating).

The first category in Bloom's taxonomy is *remembering*. For a person doing a literature review this would include tasks such as recognition, retrieval and recollection of the relevant literature. During this stage relevant books, articles, monographs, dissertations, etc. are identified and read. Bloom's second category *understanding* occurs as the scholar comprehends the material they have collected and read. This step is critical because no one can write clearly about something they do not understand. Understanding may be challenging because the literature could introduce the scholar to new terminology, conceptual framework and methodology. Comprehension (particularly for new scholars) is often improved by taking careful notes. In Bloom's third category *applying* the scholar is able to make connections between the literature and his or her larger research project. This is particularly true if the literature review is to be a chapter in a future empirical study. The literature review begins to inform the research question, and methodological approaches. When scholars *analyze* (fourth category in Bloom's taxonomy) they are able to separate material into parts and figure out how the parts fit together. Analysis of the literature allows the scholar to develop frameworks for analysis and the ability to see the big picture and know how details from the literature fit within the big picture. Analysis facilitates the development of an outline (list). The books, articles and monographs read will be of different quality and value. When scholars use Bloom's fifth category *evaluating* they are able to see the strengths and weaknesses of the theories, arguments, methodology and findings of the literature they have collected and read. When scholars engage in *creating* the final category in Bloom's taxonomy, they bring creativity to the process of doing a literature review. In other words, they draw new and original insights from the literature. They may be able to find a fresh and original research question, identify a heretofore, unknown gap in the literature or make surprising connections. By understanding how ways of thinking connect to tasks of a literature review, a scholar is able to be self-reflective and bring metacognition to the *process* of reviewing the literature.

Most of these tasks occur before the writing even begins. The *process* of reviewing the literature and writing a literature review can be complicated and lengthy. It is helpful to bring a system of organization and planning to the task. When an orderly system can be designed, it is easier to keep track of the articles, books, materials read, notes, outlines and drafts.

Systematic Review

A systematic review is a type of literature review that collects and critically analyzes multiple research studies or papers. A review of existing studies is often quicker and cheaper than embarking on a new study. Researchers use methods that are selected before one or more research questions are formulated, and then they aim to find and analyze studies that relate to and answer those questions. Systematic reviews of randomized controlled trials are key in the practice of evidence-based medicine.

An understanding of systematic reviews, and how to implement them in practice, is highly recommended for professionals involved in the delivery of health care. Besides health interventions, systematic reviews may examine clinical tests, public health interventions, environmental interventions, social interventions, adverse effects, and economic evaluations. Systematic reviews are not limited to medicine and are quite common in all other sciences where data are collected, published in the literature, and an assessment of methodological quality for a precisely defined subject would be helpful.

Characteristics

A systematic review aims to provide a complete, exhaustive summary of current literature relevant to a research question. The first step in conducting a systematic review is to perform a thorough search of the literature for relevant papers. The *Methodology* section of a systematic review will list all of the databases and citation indexes that were searched such as Web of Science, Embase, and PubMed and any individual journals that were searched. The titles and abstracts of identified articles are checked against pre-determined criteria for eligibility and relevance to form an inclusion set. This set will relate back to the research problem. Each included study may be assigned an objective assessment of methodological quality preferably by using methods conforming to the Preferred Reporting Items for Systematic Reviews and Meta-Analyses (PRISMA) statement (the current guideline) or the high quality standards of Cochrane collaboration.

Systematic reviews often, but not always, use statistical techniques (meta-analysis) to combine results of eligible studies, or at least use scoring of the levels of evidence depending on the methodology used. An additional rater may be consulted to resolve any scoring differences between raters. Systematic review is often applied in the biomedical or healthcare context, but it can be applied in any field of research. Groups like the Campbell Collaboration are promoting the use of systematic reviews in policy-making beyond just healthcare.

A systematic review uses an objective and transparent approach for research synthesis, with the aim of minimizing bias. While many systematic reviews are based on an explicit quantitative meta-analysis of available data, there are also qualitative reviews which adhere to standards for gathering, analyzing and reporting evidence. The EPPI-Centre has been influential in developing methods for combining both qualitative and quantitative research in systematic reviews. The PRISMA statement suggests a standardized way to ensure a transparent and complete reporting of systematic reviews, and is now required for this kind of research by more than 170 medical journals worldwide.

Recent developments in systematic reviews include realist reviews, and the meta-narrative approach. These approaches try to overcome the problems of methodological and epistemological heterogeneity in the diverse literatures existing on some subjects.

Stages

The main stages of a systematic review are:

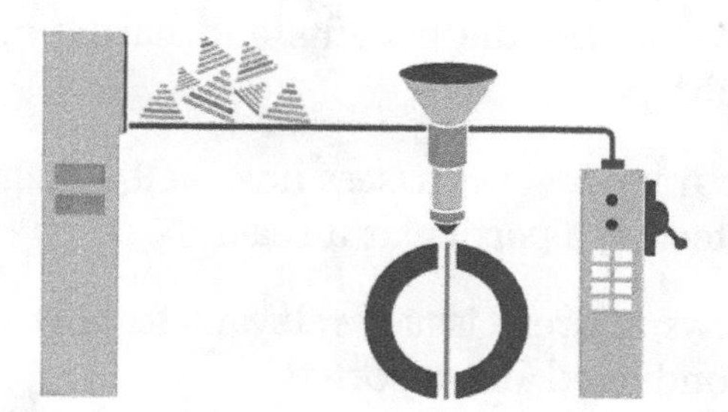

A visualisation of data being 'extracted' and 'combined' in a Cochrane systematic review.

1. Defining a question and agreeing an objective method.
2. A search for relevant data from research that matches certain criteria. For example, only selecting research that is good quality and answers the defined question.
3. 'Extraction' of relevant data. This can include how the research was done (often called the method or 'intervention'), who participated in the research (including how many people), how it was paid for (for example funding sources) and what happened (the outcomes).
4. Assess the quality of the data by judging it against criteria identified at the first stage.
5. Analyse and combine the data (using complex statistical methods) which give an overall result from all of the data. This combination of data can be visualised using a blobbogram (also called a forest plot). The diamond in the blobbogram represents the combined results of all the data included. Because this combined result uses data from more sources than just one data set, it's considered more reliable and better evidence, as the more data there is, the more confident we can be of conclusions.

Once these stages are complete, the review may be published, disseminated and translated into practice after being adopted as evidence.

Cochrane Collaboration

The Cochrane Collaboration is a group of over 31,000 specialists in healthcare who systematically review randomised trials of the effects of prevention, treatments and rehabilitation as well as health systems interventions. When appropriate, they also include the results of other types of research. Cochrane Reviews are published in *The Cochrane Database of Systematic Reviews* section of the Cochrane Library. The 2015 impact factor for *The Cochrane Database of Systematic Reviews* was 6.103, and it was ranked 12th in the "Medicine, General & Internal" category. There are six types of Cochrane Review:

1. Intervention reviews assess the benefits and harms of interventions used in healthcare and health policy.
2. Diagnostic test accuracy reviews assess how well a diagnostic test performs in diagnosing and detecting a particular disease.
3. Methodology reviews address issues relevant to how systematic reviews and clinical trials are conducted and reported.
4. Qualitative reviews synthesize qualitative and quantitative evidence to address questions on aspects other than effectiveness.
5. Prognosis reviews address the probable course or future outcome(s) of people with a health problem.
6. Overviews of Systematic Reviews (OoRs) are a new type of study in order to compile multiple evidence from systematic reviews into a single document that is accessible and useful to serve as a friendly front end for the Cochrane Collaboration with regard to healthcare decision-making.

The Cochrane Collaboration provides a handbook for systematic reviewers of interventions which "provides guidance to authors for the preparation of Cochrane Intervention reviews." The *Cochrane Handbook* outlines eight general steps for preparing a systematic review:

1. Defining the review question(s) and developing criteria for including studies
2. Searching for studies
3. Selecting studies and collecting data
4. Assessing risk of bias in included studies
5. Analysing data and undertaking meta-analyses
6. Addressing reporting biases
7. Presenting results and "summary of findings" tables
8. Interpreting results and drawing conclusions

The Cochrane Handbook forms the basis of two sets of standards for the conduct and reporting of Cochrane Intervention Reviews (MECIR - Methodological Expectations of Cochrane Intervention Reviews)

The Cochrane Collaboration logo visually represents how results from some systematic reviews can be explained. The lines within illustrate the summary results from an iconic systematic review showing the benefit of corticosteroids, which 'has probably saved thousands of premature babies'.

The Campbell Collaboration

The Campbell Collaboration is one of a number of groups promoting evidence-based policy in the social sciences. The Campbell Collaboration "helps people make well-informed decisions by preparing, maintaining and disseminating systematic reviews in education, crime and justice, social welfare and international development. It is a sister initiative of Cochrane. The Campbell Collaboration was created in 2000 and the inaugural meeting in Philadelphia, USA, attracted 85 participants from 13 countries.

Strengths and Weaknesses of Systematic Reviews

While systematic reviews are regarded as the strongest form of medical evidence, a review of 300 studies found that not all systematic reviews were equally reliable, and that their reporting can be improved by a universally agreed upon set of standards and guidelines. A further study by the same group found that of 100 systematic reviews monitored, 7% needed updating at the time of publication, another 4% within a year, and another 11% within 2 years; this figure was higher in rapidly changing fields of medicine, especially cardiovascular medicine. A 2003 study suggested that extending searches beyond major databases, perhaps into grey literature, would increase the effectiveness of reviews.

Roberts and colleagues highlighted the problems with systematic reviews, particularly those conducted by the Cochrane Collaboration, noting that published reviews are often biased, out of date and excessively long. They criticized Cochrane reviews as not being sufficiently critical in the selection of trials and including too many of low quality. They proposed several solutions, including limiting studies in meta-analyses and reviews to registered clinical trials, requiring that original data be made available for statistical checking, paying greater attention to sample size estimates, and eliminating dependence on only published data.

Some of these difficulties were noted early on as described by Altman: "much poor research arises because researchers feel compelled for career reasons to carry out research that they are ill equipped to perform, and nobody stops them." Methodological limitations of meta-analysis have also been noted. Another concern is that the methods used to conduct a systematic review are sometimes changed once researchers see the available trials they are going to include. Bloggers have described retractions of systematic reviews and published reports of studies included in published systematic reviews.

Systematic reviews are increasingly prevalent in other fields, such as international development research. Subsequently, a number of donors – most notably the UK Department for International Development (DFID) and AusAid – are focusing more attention and resources on testing the appropriateness of systematic reviews in assessing the impacts of development and humanitarian interventions.

References

- Porta, M., ed. (2008). A Dictionary of Epidemiology (5th ed.). New York: Oxford University Press. ISBN 9780195314496.
- Rothman, K. (2002). Epidemiology: An Introduction. Oxford, England: Oxford University Press. ISBN 978-0-19-513554-1.
- Rothman, K. J.; Greenland, S.; Lash, T. L. (2008). Modern Epidemiology (3rd ed.). Wolters Kluwer, Lippincott Williams & Wilkins. ISBN 978-0-7817-5564-1.
- Adams, John; Khan, Hafiz T A; Raeside, Robert (2007). Research methods for graduate business and social science students. New Delhi: SAGE Publications. p. 56. ISBN 9780761935896.
- Herman J. Ader; Gideon J. Mellenbergh; David J. Hand (2008). “Methodological quality”. Advising on Research Methods: A consultant’s companion. Johannes van Kessel Publishing. ISBN 978-90-79418-02-2.
- Lamb, David. “The Uses of Analysis: Rhetorical Analysis, Article Analysis, and the Literature Review”. Academic Writing Tutor. Archived from the original on 23 May 2014. Retrieved 26 February 2016.
- Higgins JPT; Green S (eds.). “Cochrane handbook for systematic reviews of interventions, version 5.1.0 (updated March 2011)”. The Cochrane Collaboration. Retrieved 2 June 2016.
- “Animated Storyboard: What Are Systematic Reviews?”. cccrg.cochrane.org. Cochrane Consumers and Communication. Retrieved 1 June 2016.
- The Cochrane Library. 2015 impact factor. Cochrane Database of Systematic Reviews (CDSR) Retrieved 2016-07-20.
- anonymous. “Animated Storyboard: What Are Systematic Reviews? | Cochrane Consumers and Communication”. cccrg.cochrane.org. Retrieved 2016-06-01.
- “Retraction Of Scientific Papers For Fraud Or Bias Is Just The Tip Of The Iceberg”. IFL Science!. Retrieved 29 June 2015.
- “Retraction and republication for Lancet Resp Med tracheostomy paper”. Retraction Watch. Retrieved 29 June 2015.
- “What is EBM?”. Centre for Evidence Based Medicine. 2009-11-20. Archived from the original on 2011-04-06. Retrieved 2011-06-17.

7

Research Ethics

The ethics that are involved in researches are known as research ethics. It includes topics such as ethical research in social science, clinical research ethics, committee on publication ethics and privacy for research participants. This section has been carefully written to provide an easy understanding of the varied facets of research ethics.

Research Ethics

Research ethics involves the application of fundamental ethical principles to a variety of topics involving research, including scientific research. These include the design and implementation of research involving human experimentation, animal experimentation, various aspects of academic scandal, including scientific misconduct (such as fraud, fabrication of data and plagiarism), whistleblowing; regulation of research, etc. Research ethics is most developed as a concept in medical research. The key agreement here is the 1964 Declaration of Helsinki. The Nuremberg Code is a former agreement, but with many still important notes. Research in the social sciences presents a different set of issues than those in medical research.

The academic research enterprise is built on a foundation of trust. Researchers trust that the results reported by others are sound. Society trusts that the results of research reflect an honest attempt by scientists and other researchers to describe the world accurately and without bias. But this trust will endure only if the scientific community devotes itself to exemplifying and transmitting the values associated with ethical research conduct.

There are many ethical issues to be taken into serious consideration for research. Sociologists need to be aware of having the responsibility to secure the actual permission and interests of all those involved in the study. They should not misuse any of the information discovered, and there should be a certain moral responsibility maintained towards the participants. There is a duty to protect the rights of people in the study as well as their privacy and sensitivity. The confidentiality of those involved in the observation must be carried out, keeping their anonymity and privacy secure. All of these ethics must be honoured unless there are other overriding reasons to do so - for example, any illegal or terrorist activity.

Research ethics in a medical context is dominated by principlism, an approach that has been criticised as being decontextualised. Medical research involving human experimentation is overseen by an ethics committee, in most countries working under legislation based on the Declaration of Helsinki and its later revisions.

Research ethics is different throughout different types of educational communities. Every community has its own set of morals. In Anthropology research ethics were formed to protect those who are being researched and to protect the researcher from topics or events that may be unsafe or may make either party feel uncomfortable. It is a widely observed guideline that anthropologists use especially when doing ethnographic fieldwork.

Research informants participating in individual or group interviews as well as ethnographic fieldwork are often required to sign an informed consent form which outlines the nature of the project. Informants are typically assured anonymity and will be referred to using pseudonyms. There is however growing recognition that these formal measures are insufficient and do not necessarily warrant a research project 'ethical'. Research with people should therefore not be based solely on dominant and de-contextualised understandings of ethics, but should be negotiated reflexively and through dialogue with participants as a way to bridge global and local understandings of research ethics.

In Canada, there are many different types of research ethic boards that approve applications for research projects. The most common document that Canadian Universities follow is the Tri-Council Policy Statement. However, there are other types of documents geared towards different educational aspects such as: biology, clinical practices, bio-technics and even stem cell research. The Tri-Council is actually the top three government grant agencies in Canada. If one was to do research in Canada and apply for funds, their project would have to be approved by the Tri-Council.

Furthermore, it is the researchers ethical responsibility to not harm the humans they are studying, they also have a responsibility to science, and the public, as well as to future students.

Key Issues

In terms of research publications, a number of key issues include and are not restricted to:

- Honesty. Honesty and integrity is a duty of each author and person, expert-reviewer and member of journal editorial boards.
- Review process. The peer-review process contributes to the quality control and it is an essential step to ascertain the standing and originality of the research.
- Ethical standards. Recent journal editorials presented some experience of unscrupulous activities.

- Authorship. Who may claim a right to authorship? In which order should the authors be listed?

Notable Research Experiments

From the 1950's to the 1960's, Chester M. Southam injected HeLa cancer cells into cancer patients, healthy individuals, and prison inmates at the Ohio Penitentiary. These experiments were especially controversial because many of Southam's subjects did not give informed consent to the injections. Although the prison inmates were aware that they were being injected with cancer cells, the elderly patients at the Jewish Chronic Disease Hospital and cancer patients did not. As a result of these experiments, the Board of Regents of the University of the State of New York decided to suspend Southam's medical license for one year, however, he was instead only placed on a one-year probation. Southam later went on to become the president of the American Association for Cancer Research.

Ethical Research in Social Science

Doing research in an ethical manner becomes extremely important when dealing with human subjects.

Research is the systematic process of collecting and analyzing information (data) in order to increase our understanding of the phenomenon about which we are concerned or interested and communicating what we discovered to the larger scientific community. The goal is to study ethics and what should occur in regard to human subject research treatment.

Historical Development

"HSR experiments were recorded during vaccination trials in the 1700s. In these early trials, physicians used themselves or their slaves as test subjects. Experiments on others were often conducted without informing the subjects of dangers associated with such experiments". Before the Pure Food and Drug Act was passed in 1906 there were no regulations or concerns with regard to research ethics on humans. The Nuremberg Code, in 1946, was the first law which stated that researchers must have consent from their human subjects. This was due in whole to the killings of numerous individuals within Nazi concentration camps. Research has been conducted on human subjects dating back to WWII when the Nazis experimented (unethically) on the Jews. "The Nazi physicians performed brutal medical experiments upon helpless concentration camp inmates. These acts of torture were characterized by several shocking features:

- persons were forced to become subjects in very dangerous studies against their will;

- nearly all subjects endured incredible suffering, mutilation, and indescribable pain;
- the experiments often were deliberately designed to terminate in a fatal outcome for their victims" (Cohen).

The basis of the Nuremberg Code is that the benefits of the research must outweigh the risks.

Research has been conducted unethically in other experiments, not in regard to torture but in cases of consent, deception, privacy, and confidentiality. Such experiments include: the Milgram Experiment of 1961 (electric shock treatment), Humphrey's Tearoom Trade of 1970 (male on male sexual encounters), and the Zimbardo Guard Study of 1971 (college student simulated prison experiment) just to name a few. In these experiments the subjects did not always know what they were getting into or were not all voluntarily participating.

It was not until the Institutional Review Board (IRB) was established in 1974 by the Department of Health, Education, and Welfare that these unlawful experiments against humans came to a halt.

Another step in ethics was the Tuskegee Syphilis Study (1932-1972). Six hundred African American males were closely watched for forty years. Four hundred of the six hundred had been stricken with syphilis yet were not told about their disease. The participants were denied treatment and many died during the study. Once the word of the study reached others only then was it stopped. President Clinton made a heartfelt apology to all of those affected and their families in 1997.

Institutional Review Board

An Institutional Review Board is a group of organizational and community representatives required by federal law to review the ethical issues in all proposed research that is federally funded, involves human subjects, or has any potential for harm to subjects (Schutt I-18). Federal regulations require that every institution that seeks federal funding for biomedical or behavioral research on human subjects have an institutional review board that reviews research proposals.

A series of recommendations focuses on improving ethics review of protocols...

Recommendation: The Institutional Review Board (IRB), as the principal representative of the interests of potential research participants, should focus its full committee deliberations and oversight primarily on the ethical aspects of protection issues. To reflect this role, IRBs should be appropriately renamed within research organizations' internal documents that define institutional structure and polices. The committee suggests the name "Research Ethics Review Board" (Research ERB). *(Recommendation 3.1)*

To address such needs most institutions and organizations have formulated an Institutional Review Board (IRB), a panel of persons who reviews grant proposals with respect to ethical implications and decides whether additional actions need to be taken to assure the safety and rights of participants. By reviewing proposals for research, IRBs also help to protect both the organization and the researcher against potential legal implications of neglecting to address important ethical issues of participants.

Human Subject Research/Ethical Principles

Human Subject Research - Ethical Principles

The Code's standards concerning the treatment of human subjects include federal regulations and ethics guidelines emphasized by most professional science organizations:

Research should cause no harm to subjects, at no point should subjects feel distressed. Deception needs to be left out of the research process, under no circumstance should a researcher lie to their subjects. Participation in research should be voluntary, and therefore subjects must give their informed consent to participate in the research. Researchers should be very cautious when dealing with vulnerable clients (persons who are mentally ill, incarcerated people, or minors) they should make sure to get the proper consent. Researchers should fully disclose their identity. Anonymity or confidentiality must be maintained for individual research participants unless it is voluntary and explicitly waived. Actively attempt to remove from the research records any elements that might indicate the subjects identities. And finally, benefits from a research project should outweigh any foreseeable risks.

The specific structure of a protection program is secondary to its performance of several essential functions. These functions include:

1. Comprehensive review of protocols, (including scientific, financial conflict of interest, and ethical reviews),
2. Ethically sound participant-investigator interactions,
3. Ongoing (and risk appropriate) safety monitoring throughout the conduct of the study,
4. Quality improvement (QI) and compliance activities.

Some principles Jones states in regard to electronic information gathering principles follows:

1. Openness: Existence of data banks should be publicly known.
2. Individual Access and Correction: People should have access to the data collected about themselves.

3. Collection Limitation and Relevance: Personal data should be collected for one specific, legitimate purpose.
4. Use Limitations: Information should be used only for purposes specified at the time of collection.
5. Disclosure Limitation: Personal data is not to be communicated externally without consent of the subject who supplied the data.
6. Security: Personal data should be reasonably guarded against risks such as loss, unauthorized access, modification, or disclosure.

Ethical Concerns

Researchers have ethical obligations to take into account when conducting interviews (one-on-one or face-to-face), case studies (individual, group, or event), focus groups (6-10 people), unobtrusive measures (artifacts, things left behind), histography (follows a person's life history), or observations (ethnography). Researchers must use a systematic process to collect data without interfering or harming your subjects.

List of Ethical concerns to take into consideration when performing research on human subjects:

- privacy: a researcher should never breach a research subject's privacy;
- misrepresentation: researchers should not hide potential conflicts of interest, or mislead subjects as to the nature of the research;
- researchers should not harm or distress (physically or psychologically) their research subjects. Throughout their research process they must take all steps necessary to ensure their personal biases or preconceptions to not influence the conduct or findings of the research;
- researchers should never ever put their subjects in a compromising position where there is a potential for danger.

As a relatively new form of interpersonal contact minus nonvisual cues, within internet circles, the need has arisen to develop a "netiquette", or rules of thumb, to encourage politeness, civility, and enhanced understanding among participants. These codes have been formulated by users as communicative problems are encountered and identified in the process of employing the medium.

Current Research/New Directions in Research

Although formation of such rules of conduct and ethical inquiries into the communicative use of the computer are still at an early stage of development, analogous guidelines for encouraging ethical practices in the conduct of on-line research are only now being discussed. At present, because the internet is part of a technology in continual flux and rapid evolution, there is incomplete awareness of the issues at stake, let alone consensus on the best ways to proceed. Nonetheless, several key ethical concerns have become apparent. Perhaps chief among these issues of:

- Privacy
- Confidentiality
- Informed consent
- Appropriation of others' personal stories.

Clinical Research Ethics

Clinical research ethics are the set of relevant ethics considered in the conduct of a clinical trial in the field of clinical research. It borrows from the broader fields of research ethics and medical ethics.

Governance

Most directly a local institutional review board oversees the clinical research ethics of any given clinical trial. The institutional review board understands and acts according to local and national law. Each countries national law is guided by international principles, such as the Belmont Report's directive that all study participants have a right to "respect for persons", "beneficence", and "justice" when participating in clinical research.

Study Participant Rights

Participants in clinical research have rights which they should expect, including the following:

- right to Informed consent
- Shared decision-making
- Privacy for research participants
- Return of results
- Right to withdraw

Vulnerable Populations

There is a range of autonomy which study participants may have in deciding their participation in clinical research. Researchers refer to populations which have low autonomy as "vulnerable populations"; these are groups which may not be able to fairly decide for themselves whether to participate in clinical trials. Examples of groups which are vulnerable populations include incarcerated persons, children, prisoners, soldiers, people under detention, migrants, persons exhibiting insanity or any other condition which precludes their autonomy, and to a lesser extent, any population for which there is reason to believe that the research study could seem particularly or unfairly persuasive or misleading. There are particular ethical problems using children in clinical trials.

Committee on Publication Ethics

The Committee on Publication Ethics (COPE) is a nonprofit organization whose mission is to define best practice in the ethics of scholarly publishing and to assist editors, publishers, etc. to achieve this.

Mission

COPE provides advice to editors and publishers on all aspects of publication ethics and, in particular, how to handle cases of research and publication misconduct. It also provides a forum for its members to discuss individual cases (meeting four times a year in the UK and once a year in North America). COPE does not investigate individual cases but encourages editors to ensure that cases are investigated by the appropriate authorities (usually a research institution or employer).

COPE also funds research on publication ethics, publishes a quarterly newsletter and organises annual seminars in the United Kingdom and the United States. Besides, COPE has created an audit tool for members to measure compliance with its *Code of Conduct and Best Practice Guidelines*.

History

COPE was established in 1997 by a small group of medical journal editors in the United Kingdom. Now it has over 6000 members worldwide, from all academic fields. Paid membership is open to editors of academic journals and others interested in publication ethics, and varies from £50 to £42,028 per year depending on the membership type.

COPE's first guidelines were developed after discussion at the COPE meeting in April 1999 and were published as *Guidelines on Good Publication Practice* in the Annual Report in 1999. On their basis, in 2004, the first edition of *Code of Conduct for Editors* was drafted. The draft underwent wide consultation with COPE Council as well as editors and publishers. The Code was published on the first COPE website in November 2004, with an Editorial in the BMJ. The Code was later revised and improved, and other guidance was also issued to aid editors and publishers in the fight against research and publication misconduct (scientific misconduct).

Structure

COPE is run by a governing Council whose members are trustees of the charity. COPE has an independent Ombudsman to adjudicate disputes between COPE members or between them and the organisation. The day-to-day running of COPE is handled by the Operations Manager and Administrator.

Links with Other Organizations

It also has links with the Council of Science Editors, the European Association of Science Editors, the International Society of Managing and Technical Editors and the World Association of Medical Editors.

Privacy for Research Participants

Privacy for research participants is a concept in research ethics which states that a person in human subject research has a right to privacy when participating in research. Some examples of typical scenarios are that a surveyor doing social research conducts an interview where the research participant is a respondent, or when a researcher for a clinical trial asks for a blood sample from a participant to see if there is a relationship between something which can be measured in blood and a person's health. In both cases, the ideal outcome is that any participant can join the study and neither the researcher nor the study design nor the publication of the study results would ever identify any participant in the study.

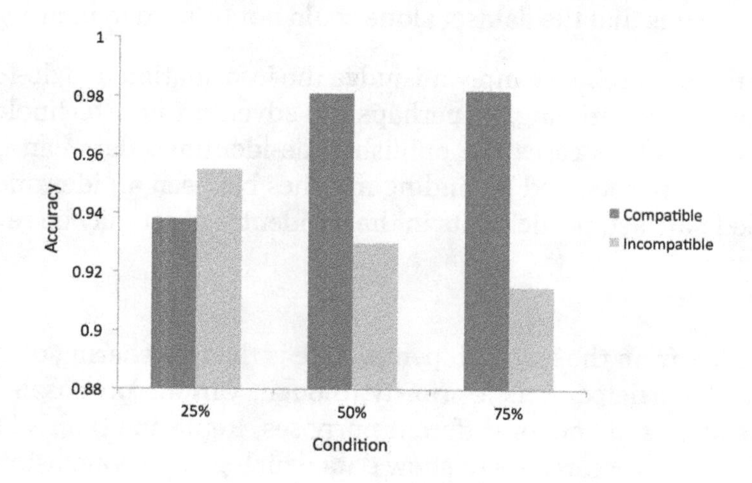

People decide to participate in research for any number of different reasons, such as a personal interest, a desire to promote research which benefits their community, or for other reasons. Various guidelines for human subject research protect study participants who choose to participate in research, and the international consensus is that the rights of people who participate in studies are best protected when the study participant can trust that researchers will not connect the identities of study participants with their input into the study.

Many study participants have experienced problems when their privacy was not upheld after participating in research. Sometimes privacy is not kept because of insufficient study protection, but also sometimes it is because of unanticipated problems with the study design which inadvertently compromise privacy. The privacy of research participants is typically protected by the research organizer, but the institutional review board is a designated overseer which monitors the organizer to provide protection to study participants.

Information Privacy

Researchers publish data that they get from participants. To preserve participants' privacy, the data goes through a process to de-identify it. The goal of such a process would be to remove protected health information which could be used to connect a study participant to their contribution to a research project.

Privacy Attacks

A privacy attack is the exploitation of an opportunity of someone to identify a study participant based on public research data. The way that this might work is that researchers collect data, including confidential identifying data, from study participants. This produces an identified dataset. Before the data is sent for research processing, it is "de-identified", which means that personally identifying data is removed from the dataset. Ideally this means that the dataset alone could not be used to identify a participant.

In some cases the researchers simply misjudge the information in a de-identified dataset and actually it is identifying, or perhaps the advent of new technology makes the data identifying. In other cases the published de-identified data can be cross-referenced with other data sets, and by finding matches between an identified dataset and the de-identified data set, participants in the de-identified set may be revealed.

Risk Mitigation

The ideal situation from the research perspective is the free sharing of data. Since privacy for research participants is a priority, though, various proposals for protecting participants have been made for different purposes. Replacing the real data with synthetic data allows the researchers to show data which gives a conclusion equivalent to the one drawn by the researchers, but the data may have problems such as being unfit for repurposing for other research. Other strategies include "noise addition" by making random value changes or "data swapping" by exchanging values across entries. Still another approach is to separate the identifiable variables in the data from the rest, aggregate the identifiable variables and reattach them with the rest of the data. This principle has been used successfully in creating maps of diabetes in Australia and the United Kingdom using confidential General Practice clinic data.

Biobank Privacy

A biobank is a place where human biological specimens are kept for research, and often where genomics data is paired with phenotype data and personally-identifying data. For many reasons, biobank research has created new controversies, perspectives, and challenges for satisfying the rights of student participants and the needs of the researchers to access resources for their work.

One problem is that if even a small percentage of genetic information is available, that information can be used to uniquely identify the individual from which it came. Studies have shown that a determination of whether an individual participated in a study can be made even from reporting of aggregate data.

Negative Consequences

When research participants have their identities revealed they may have problems. Concerns include facing genetic discrimination from an insurance company or employer. Respondents in the United States have expressed a desire to have their research data to be restricted from access by law enforcement agencies, and would want to prevent a connection between study participation and legal consequences of the same. Another fear study participants have is about the research revealing private personal practices which a person may not want to discuss, such as a medical history which includes a sexually transmitted disease, substance abuse, psychiatric treatment, or an elective abortion. In the case of genomic studies on families, genetic screening may reveal that paternity is different from what had been supposed. For no particular reason, some people may find that if their private information becomes disclosed because of research participation, they may feel invaded and find the entire system distasteful.

Privacy Controversies

- Netflix Prize - researchers release a database with approximate years of birth, zip codes, and movie-watching preferences. Other researchers say that based even on this limited information, many people can be identified and their movie preferences could be discovered. People objected to having their movie-watching habits become publicly known.

- Tearoom Trade - a university researcher published information revealing persons who engaged in illicit sex, and research participants did not consent to be identified.

References

- "COPE Individual and Corporate Subscription rates 2011" (PDF). Committee on Publication Ethics (COPE). 2011. Retrieved February 22, 2013.
- Beecher, Henry K. (June 16, 1966). "Ethics and Clinical Research". N Engl J Med. 274 (24): 1354–1360. doi:10.1056/NEJM196606162742405. PMID 5327352. Retrieved 11 January 2012.
- Skloot, Rebecca (2010). The Immortal Life of Henrietta Lacks. New York: Broadway Paperbacks. p. 127-130.

8

Understanding Methodology

Methodology is an organized analysis of the methods that are applied to a particular field of study. It mainly consists of concepts such as paradigm, theoretical model and quantitative techniques. The other topics explained are art methodology and rationalism. In order to completely understand methodology, it is necessary to understand the themes elucidated in the following section.

Methodology

Methodology is the systematic, theoretical analysis of the methods applied to a field of study. It comprises the theoretical analysis of the body of methods and principles associated with a branch of knowledge. Typically, it encompasses concepts such as paradigm, theoretical model, phases and quantitative or qualitative techniques.

A methodology does not set out to provide solutions - it is, therefore, not the same as a method. Instead, a methodology offers the theoretical underpinning for understanding which method, set of methods, or best practices can be applied to specific case, for example, to calculate a specific result.

It has been defined also as follows:

1. "the analysis of the principles of methods, rules, and postulates employed by a discipline";
2. "the systematic study of methods that are, can be, or have been applied within a discipline";
3. "the study or description of methods".

Relationship between Methodology, Theory, Paradigm, Algorithm and Method

The *methodology* is the general research strategy that outlines the way in which research is to be undertaken and, among other things, identifies the methods to be used in it. These *methods*, described in the methodology, define the means or modes of data collection or, sometimes, how a specific result is to be calculated. *Methodology* does not define specific methods, even though much attention is given to the nature and kinds of processes to be followed in a particular procedure or to attain an objective.

When proper to a study of methodology, such processes constitute a *constructive generic framework*, and may therefore be broken down into sub-processes, combined, or their sequence changed.

A *paradigm* is similar to a methodology in that it is also a *constructive framework*. In theoretical work, the development of paradigms satisfies most or all of the criteria for methodology. An *algorithm*, like a paradigm, is also a type of *constructive framework*, meaning that the construction is a logical, rather than a physical, array of connected elements.

Any description of a means of calculation of a specific result is always a description of a method and never a description of a methodology. It is thus important to avoid using *methodology* as a synonym for *method* or *body of methods*. Doing this shifts it away from its true epistemological meaning and reduces it to being the procedure itself, or the set of tools, or the instruments that should have been its outcome. A methodology is the design process for carrying out research or the development of a procedure and is not in itself an instrument, or method, or procedure for doing things.

Methodology and *method* are not interchangeable. In recent years however, there has been a tendency to use *methodology* as a "pretentious substitute for the word *method*". Using *methodology* as a synonym for *method* or *set of methods* leads to confusion and misinterpretation and undermines the proper analysis that should go into designing research.

Art Methodology

Art methodology refers to a *studied* and constantly reassessed, questioned method within the arts, as opposed to a method merely applied (without thought). This process of studying the method and reassessing its effectiveness allows art to move on and change. It is not the thing itself but it is an essential part of the process.

An artist drawing, for instance, may choose to draw from what he or she observes in front of them, or from what they *imagine* or from what they already know about the subject. These 3 methods will, very probably, produce 3 very different pictures. A careful methodology would include examination of the materials and tools used and how a different type of canvas/brush/paper/pencil/rag/camera/chisel etc. would produce a different effect. The artist may also look at various effects achieved by starting in one part of a canvas first, or by working over the whole surface equally. An author may experiment with stream of consciousness writing, as opposed to naturalistic narrative, or a combination of styles.

Fine Art Compared with Traditional Crafts

In stark contrast to fine art practice is the traditional craft form. With traditional crafts, the method is handed down from generation to generation with often very little change in techniques. It is usually fair to say that folk crafts employ a method but not an art methodology, since that would involve rigorous questioning and criticising of the tradition.

Art Methodology Compared with Science Methodology

An art methodology differs from a science methodology, perhaps mainly insofar as the artist is not always after the same goal as the scientist. In art it is not necessarily all about establishing the exact truth so much as making the most effective form (painting, drawing, poem, novel, performance, sculpture, video, etc.) through which ideas, feelings, perceptions can be communicated to a public. With this purpose in mind, some artists will exhibit preliminary sketches and notes which were part of the process leading to the creation of a work. Sometimes, in Conceptual art, the preliminary process is the only part of the work which *is* exhibited, with no visible end result displayed. In such a case the *"journey"* is being presented as more important than the destination. Conceptual artist Robert Barry once put on an exhibition where the door of the gallery remained shut and a sign on the door informed visitors that the gallery would be closed for the exhibition. These kind of works question accepted concepts, such as that of having a tangible work of art as end result.

Some Art Methodology Statements

Global Responsibility

The Peace Through Art methodology developed by the International Child Art Foundation (ICAF) was recognized as a Stockholm Challenge Finalist for best practice and innovative initiative, in Stockholm, Sweden. ICAF's statement on the methodology of the programme says: "the Peace Through Art methodology draws upon the creativity and imagination of young people, and teaches them the ethics of responsibility in this interdependent global village that has come to be our world. The methodology incorporates best practices from the fields of psychology, conflict resolution and peace education, while employing the power of the arts for self-expression, healing and communication."

Art Therapy

In Art Therapy and other applied forms of art to the needs of groups within the community art methodology is built up in relation to other methodologies, such as that of psychotherapy.

Pauline Mottram SRAsTh, gave a short paper entitled *"Towards developing a methodology to evaluate outcomes of Art Therapy in Adult Mental Illness."* for the October 2000 TAoAT Conference. Here's an excerpt:

"In addressing the effectiveness challenge and in seeking a means of measuring out-

comes of the art therapy service that I deliver, I have sought a methodology that can provide a valid and reliable quantitative outcome, whilst still respecting the aesthetic and humanistic nature of art therapy practice. This endeavour is made all the more complex by the fact that art therapy lacks a fully developed theory and it has a fragile research base. Generally it is argued that art therapy is more compatible with qualitative research designs that encompass subjectivity, rather than quantitative objective methods. Kaplan (1998, p95) states that 'Qualitative is exploratory and theory building. Quantitative tests hypothesis in order to refine and validate theory.' She holds that art therapy cannot afford to reject either form of inquiry."

Generative Art

In *"The Methodology of Generative Art"* by Tjark Ihmels and Julia Riedel, an online article at Media Art Net Mozart's "musical game of dice" is cited as a precedent for the methodology of Generative art which, say the authors, has *"established itself in nearly every area of artistic practice (music, literature , the fine arts)."*:

"Mozart composed 176 bars of music, from which sixteen were chosen from a list using dice, which then produced a new piece when performed on a piano. Sixteen bars, each with eleven possibilities, can result in 1,116 unique pieces of music. Using this historical example, the methodology of generative art can be appropriately described as the rigorous application of predefined principles of action for the intentional exclusion of, or substitution for, individual aesthetical decisions that sets in motion the generation of new artistic content out of material provided for that purpose."

Rationalism

In epistemology, rationalism is the view that "regards reason as the chief source and test of knowledge" or "any view appealing to reason as a source of knowledge or justification". More formally, rationalism is defined as a methodology or a theory "in which the criterion of the truth is not sensory but intellectual and deductive".

In an old controversy, rationalism was opposed to empiricism, where the rationalists believed that reality has an intrinsically logical structure. Because of this, the rationalists argued that certain truths exist and that the intellect can directly grasp these truths. That is to say, rationalists asserted that certain rational principles exist in logic, mathematics, ethics, and metaphysics that are so fundamentally true that denying them causes one to fall into contradiction. The rationalists had such a high confidence in reason that empirical proof and physical evidence were regarded as unnecessary to ascertain certain truths – in other words, "there are significant ways in which our concepts and knowledge are gained independently of sense experience".

Different degrees of emphasis on this method or theory lead to a range of rationalist standpoints, from the moderate position "that reason has precedence over other ways of acquiring knowledge" to the more extreme position that reason is "the unique path to knowledge". Given a pre-modern understanding of reason, rationalism is identical to philosophy, the Socratic life of inquiry, or the zetetic (skeptical) clear interpretation of authority (open to the underlying or essential cause of things as they appear to our sense of certainty). In recent decades, Leo Strauss sought to revive "Classical Political Rationalism" as a discipline that understands the task of reasoning, not as foundational, but as maieutic.

In politics, Rationalism, since the Enlightenment, historically emphasized a "politics of reason" centered upon rational choice, utilitarianism, secularism, and irreligion – the latter aspect's antitheism later ameliorated by utilitarian adoption of pluralistic rationalist methods practicable regardless of religious or irreligious ideology.

In this regard, the philosopher John Cottingham noted how rationalism, a methodology, became socially conflated with atheism, a worldview:

In the past, particularly in the 17th and 18th centuries, the term 'rationalist' was often used to refer to free thinkers of an anti-clerical and anti-religious outlook, and for a time the word acquired a distinctly pejorative force (thus in 1670 Sanderson spoke disparagingly of 'a mere rationalist, that is to say in plain English an atheist of the late edition...'). The use of the label 'rationalist' to characterize a world outlook which has no place for the supernatural is becoming less popular today; terms like 'humanist' or 'materialist' seem largely to have taken its place. But the old usage still survives.

Philosophical Usage

Rationalism is often contrasted with empiricism. Taken very broadly these views are not mutually exclusive, since a philosopher can be both rationalist and empiricist. Taken to extremes, the empiricist view holds that all ideas come to us *a posteriori*, that is to say, through experience; either through the external senses or through such inner sensations as pain and gratification. The empiricist essentially believes that knowledge is based on or derived directly from experience. The rationalist believes we come to knowledge *a priori* – through the use of logic – and is thus independent of sensory experience. In other words, as Galen Strawson once wrote, "you can see that it is true just lying on your couch. You don't have to get up off your couch and go outside and examine the way things are in the physical world. You don't have to do any science." Between both philosophies, the issue at hand is the fundamental source of human knowledge and the proper techniques for verifying what we think we know. Whereas both philosophies are under the umbrella of epistemology, their argument lies in the understanding of the warrant, which is under the wider epistemic umbrella of the theory of justification.

Theory of Justification

The theory of justification is the part of epistemology that attempts to understand the justification of propositions and beliefs. Epistemologists are concerned with various epistemic features of belief, which include the ideas of justification, warrant, rationality, and probability. Of these four terms, the term that has been most widely used and discussed by the early 21st century is "warrant". Loosely speaking, justification is the reason that someone (probably) holds a belief.

If "A" makes a claim, and "B" then casts doubt on it, "A"'s next move would normally be to provide justification. The precise method one uses to provide justification is where the lines are drawn between rationalism and empiricism (among other philosophical views). Much of the debate in these fields are focused on analyzing the nature of knowledge and how it relates to connected notions such as truth, belief, and justification.

Theses of Rationalism

At its core, rationalism consists of three basic claims. For one to consider themselves a rationalist, they must adopt at least one of these three claims: The Intuition/Deduction Thesis, The Innate Knowledge Thesis, or The Innate Concept Thesis. In addition, rationalists can choose to adopt the claims of Indispensability of Reason and or the Superiority of Reason – although one can be a rationalist without adopting either thesis.

The Intuition/Deduction Thesis

Rationale: *"Some propositions in a particular subject area, S, are knowable by us by intuition alone; still others are knowable by being deduced from intuited propositions."*

Generally speaking, intuition is *a priori* knowledge or experiential belief characterized by its immediacy; a form of rational insight. We simply just "see" something in such a way as to give us a warranted belief. Beyond that, the nature of intuition is hotly debated.

In the same way, generally speaking, deduction is the process of reasoning from one or more general premises to reach a logically certain conclusion. Using valid arguments, we can deduce from intuited premises.

For example, when we combine both concepts, we can intuit that the number three is prime and that it is greater than two. We then deduce from this knowledge that there is a prime number greater than two. Thus, it can be said that intuition and deduction combined to provide us with *a priori* knowledge – we gained this knowledge independently of sense experience.

Empiricists such as David Hume have been willing to accept this thesis for describing the relationships among our own concepts. In this sense, empiricists argue that we are allowed to intuit and deduce truths from knowledge that has been obtained *a posteriori*.

By injecting different subjects into the Intuition/Deduction thesis, we are able to generate different arguments. Most rationalists agree mathematics is knowable by applying the intuition and deduction. Some go further to include ethical truths into the category of things knowable by intuition and deduction. Furthermore, some rationalists also claim metaphysics is knowable in this thesis.

In addition to different subjects, rationalists sometimes vary the strength of their claims by adjusting their understanding of the warrant. Some rationalists understand warranted beliefs to be beyond even the slightest doubt; others are more conservative and understand the warrant to be belief beyond a reasonable doubt.

Rationalists also have different understanding and claims involving the connection between intuition and truth. Some rationalists claim that intuition is infallible and that anything we intuit to be true is as such. More contemporary rationalists accept that intuition is not always a source of certain knowledge – thus allowing for the possibility of a deceiver who might cause the rationalist to intuit a false proposition in the same way a third party could cause the rationalist to have perceptions of nonexistent objects.

Naturally, the more subjects the rationalists claim to be knowable by the Intuition/Deduction thesis, the more certain they are of their warranted beliefs, and the more strictly they adhere to the infallibility of intuition, the more controversial their truths or claims and the more radical their rationalism.

To argue in favor of this thesis, Gottfried Wilhelm Leibniz, a prominent German philosopher, says, "The senses, although they are necessary for all our actual knowledge, are not sufficient to give us the whole of it, since the senses never give anything but instances, that is to say particular or individual truths. Now all the instances which confirm a general truth, however numerous they may be, are not sufficient to establish the universal necessity of this same truth, for it does not follow that what happened before will happen in the same way again. ... From which it appears that necessary truths, such as we find in pure mathematics, and particularly in arithmetic and geometry, must have principles whose proof does not depend on instances, nor consequently on the testimony of the senses, although without the senses it would never have occurred to us to think of them..."

The Innate Knowledge Thesis

Rationale: *"We have knowledge of some truths in a particular subject area, S, as part of our rational nature."*

The Innate Knowledge thesis is similar to the Intuition/Deduction thesis in the regard that both theses claim knowledge is gained *a priori*. The two theses go their separate ways when describing how that knowledge is gained. As the name, and the rationale, suggests, the Innate Knowledge thesis claims knowledge is simply part of our rational nature. Experiences can trigger a process that allows this knowledge to come into our consciousness, but the experiences don't provide us with the knowledge itself. The knowledge has been with us since the beginning and the experience simply brought into focus, in the same way a photographer can bring the background of a picture into focus by changing the aperture of the lens. The background was always there, just not in focus.

This thesis targets a problem with the nature of inquiry originally postulated by Plato in *Meno*. Here, Plato asks about inquiry; how do we gain knowledge of a theorem in geometry? We inquire into the matter. Yet, knowledge by inquiry seems impossible. In other words, "If we already have the knowledge, there is no place for inquiry. If we lack the knowledge, we don't know what we are seeking and cannot recognize it when we find it. Either way we cannot gain knowledge of the theorem by inquiry. Yet, we do know some theorems." The Innate Knowledge thesis offers a solution to this paradox. By claiming that knowledge is already with us, either consciously or unconsciously, a rationalist claims we don't really "learn" things in the traditional usage of the word, but rather that we simply bring to light what we already know.

The Innate Concept Thesis

Rationale: *"We have some of the concepts we employ in a particular subject area, S, as part of our rational nature."*

Similar to the Innate Knowledge thesis, the Innate Concept thesis suggests that some concepts are simply part of our rational nature. These concepts are *a priori* in nature and sense experience is irrelevant to determining the nature of these concepts (though, sense experience can help bring the concepts to our conscious mind).

Some philosophers, such as John Locke (who is considered one of the most influential thinkers of the Enlightenment and an empiricist) argue that the Innate Knowledge thesis and the Innate Concept thesis are the same. Other philosophers, such as Peter Carruthers, argue that the two theses are distinct from one another. As with the other theses covered under rationalisms' umbrella, the more types and greater number of concepts a philosopher claims to be innate, the more controversial and radical their position; "the more a concept seems removed from experience and the mental operations we can perform on experience the more plausibly it may be claimed to be innate. Since we do not experience perfect triangles but do experience pains, our concept of the former is a more promising candidate for being innate than our concept of the latter.

In his book, *Meditations on First Philosophy*, René Descartes postulates three classifications for our ideas when he says, "Among my ideas, some appear to be innate, some to be adventitious, and others to have been invented by me. My understanding of what a thing is, what truth is, and what thought is, seems to derive simply from my own nature. But my hearing a noise, as I do now, or seeing the sun, or feeling the fire, comes from things which are located outside me, or so I have hitherto judged. Lastly, sirens, hippogriffs and the like are my own invention."

Adventitious ideas are those concepts that we gain through sense experiences, ideas such as the sensation of heat, because they originate from outside sources; transmitting their own likeness rather than something else and something you simply cannot will away. Ideas invented by us, such as those found in mythology, legends, and fairy tales are created by us from other ideas we possess. Lastly, innate ideas, such as our ideas of perfection, are those ideas we have as a result of mental processes that are beyond what experience can directly or indirectly provide.

Gottfried Wilhelm Leibniz defends the idea of innate concepts by suggesting the mind plays a role in determining the nature of concepts, to explain this, he likens the mind to a block of marble in the *New Essays on Human Understanding*, "This is why I have taken as an illustration a block of veined marble, rather than a wholly uniform block or blank tablets, that is to say what is called tabula rasa in the language of the philosophers. For if the soul were like those blank tablets, truths would be in us in the same way as the figure of Hercules is in a block of marble, when the marble is completely indifferent whether it receives this or some other figure. But if there were veins in the stone which marked out the figure of Hercules rather than other figures, this stone would be more determined thereto, and Hercules would be as it were in some manner innate in it, although labour would be needed to uncover the veins, and to clear them by polishing, and by cutting away what prevents them from appearing. It is in this way that ideas and truths are innate in us, like natural inclinations and dispositions, natural habits or potentialities, and not like activities, although these potentialities are always accompanied by some activities which correspond to them, though they are often imperceptible."

The Other Two theses

The three aforementioned theses of Intuition/Deduction, Innate Knowledge, and Innate Concept are the cornerstones of rationalism. To be considered a rationalist, one must adopt at least one of those three claims. The following two theses are traditionally adopted by rationalists, but they aren't essential to the rationalist's position.

The Indispensability of Reason Thesis has the following rationale, "The knowledge we gain in subject area, *S*, by intuition and deduction, as well as the ideas and instances of knowledge in *S* that are innate to us, could not have been gained by us through sense experience." In short, this thesis claims that experience cannot provide what we gain from reason.

The Superiority of Reason Thesis has the following rationale, '"The knowledge we gain in subject area *S* by intuition and deduction or have innately is superior to any knowledge gained by sense experience". In other words, this thesis claims reason is superior to experience as a source for knowledge.

In addition to the following claims, rationalists often adopt similar stances on other aspects of philosophy. Most rationalists reject skepticism for the areas of knowledge they claim are knowable *a priori*. Naturally, when you claim some truths are innately known to us, one must reject skepticism in relation to those truths. Especially for rationalists who adopt the Intuition/Deduction thesis, the idea of epistemic foundationalism tends to crop up. This is the view that we know some truths without basing our belief in them on any others and that we then use this foundational knowledge to know more truths.

Background

Rationalism - as an appeal to human reason as a way of obtaining knowledge - has a philosophical history dating from antiquity. The analytical nature of much of philosophical enquiry, the awareness of apparently a priori domains of knowledge such as mathematics, combined with the emphasis of obtaining knowledge through the use of rational faculties (commonly rejecting, for example, direct revelation) have made rationalist themes very prevalent in the history of philosophy.

Since the Enlightenment, rationalism is usually associated with the introduction of mathematical methods into philosophy as seen in the works of Descartes, Leibniz, and Spinoza. This is commonly called continental rationalism, because it was predominant in the continental schools of Europe, whereas in Britain empiricism dominated.

Even then, the distinction between rationalists and empiricists was drawn at a later period and would not have been recognized by the philosophers involved. Also, the distinction between the two philosophies is not as clear-cut as is sometimes suggested; for example, Descartes and Locke have similar views about the nature of human ideas.

Proponents of some varieties of rationalism argue that, starting with foundational basic principles, like the axioms of geometry, one could deductively derive the rest of all possible knowledge. The philosophers who held this view most clearly were Baruch Spinoza and Gottfried Leibniz, whose attempts to grapple with the epistemological and metaphysical problems raised by Descartes led to a development of the fundamental approach of rationalism. Both Spinoza and Leibniz asserted that, *in principle*, all knowledge, including scientific knowledge, could be gained through the use of reason alone, though they both observed that this was not possible *in practice* for human beings except in specific areas such as mathematics. On the other hand, Leibniz admitted in his book *Monadology* that "we are all mere Empirics in three fourths of our actions."

History

Rationalist Philosophy from Antiquity

Because of the complicated nature of rationalist thinking, the nature of philosophy, and the understanding that humans are aware of knowledge available only through the use of rational thought, many of the great philosophers from antiquity laid down the foundation for rationalism though they themselves weren't rationalists as we understand the concept today.

Pythagoras (570–495 BCE)

Pythagoras was one of the first Western philosophers to stress rationalist insight. He is often revered as a great mathematician, mystic and scientist, but he is best known for the Pythagorean theorem, which bears his name, and for discovering the mathematical relationship between the length of strings on lute bear and the pitches of the notes. Pythagoras "believed these harmonies reflected the ultimate nature of reality. He summed up the implied metaphysical rationalism in the words "All is number". It is probable that he had caught the rationalist's vision, later seen by Galileo (1564–1642), of a world governed throughout by mathematically formulable laws". It has been said that he was the first man to call himself a philosopher, or lover of wisdom,

Plato (427–347 BCE)

Plato also held rational insight to a very high standard, as is seen in his works such as Meno and The Republic. Plato taught on the Theory of Forms (or the Theory of Ideas) which asserts that non-material abstract (but substantial) forms (or ideas), and not the material world of change known to us through sensation, possess the highest and most fundamental kind of reality. Plato's forms are accessible only to reason and not to sense. In fact, it is said that Plato admired reason, especially in geometry, so highly that he had the phrase "Let no one ignorant of geometry enter" inscribed over the door to his academy.

Aristotle (384–322 BCE)

Aristotle has a process of reasoning similar to that of Plato's, though he ultimately disagreed with the specifics of Plato's forms. Aristotle's great contribution to rationalist thinking comes from his use of syllogistic logic. Aristotle defines syllogism as "a discourse in which certain (specific) things having been supposed, something different from the things supposed results of necessity because these things are so." Despite this very general definition, Aristotle limits himself to categorical syllogisms which consist of three categorical propositions in his work *Prior Analytics*. These included categorical modal syllogisms.

Post-Aristotle

Though the three great Greek philosophers disagreed with one another on specific points, they all agreed that rational thought could bring to light knowledge that was self-evident – information that humans otherwise couldn't know without the use of reason. After Aristotle's death, Western rationalistic thought was generally characterized by its application to theology, such as in the works of the Islamic philosopher Avicenna and Jewish philosopher and theologian Maimonides. One notable event in the Western timelime was the philosophy of St. Thomas Aquinas who attempted to merge Greek rationalism and Christian revelation in the thirteenth-century.

Classical Rationalism

René Descartes (1596–1650)

Descartes was the first of the modern rationalists and has been dubbed the 'Father of Modern Philosophy.' Much subsequent Western philosophy is a response to his writings, which are studied closely to this day.

Descartes thought that only knowledge of eternal truths – including the truths of mathematics, and the epistemological and metaphysical foundations of the sciences – could be attained by reason alone; other knowledge, the knowledge of physics, required experience of the world, aided by the scientific method. He also argued that although dreams appear as real as sense experience, these dreams cannot provide persons with knowledge. Also, since conscious sense experience can be the cause of illusions, then sense experience itself can be doubtable. As a result, Descartes deduced that a rational pursuit of truth should doubt every belief about reality. He elaborated these beliefs in such works as *Discourse on Method*, *Meditations on First Philosophy*, and *Principles of Philosophy*. Descartes developed a method to attain truths according to which nothing that cannot be recognised by the intellect (or reason) can be classified as knowledge. These truths are gained "without any sensory experience," according to Descartes. Truths that are attained by reason are broken down into elements that intuition can grasp, which, through a purely deductive process, will result in clear truths about reality.

Descartes therefore argued, as a result of his method, that reason alone determined knowledge, and that this could be done independently of the senses. For instance, his famous dictum, *cogito ergo sum* or "I think, therefore I am", is a conclusion reached *a priori* i.e., prior to any kind of experience on the matter. The simple meaning is that doubting one's existence, in and of itself, proves that an "I" exists to do the thinking. In other words, doubting one's own doubting is absurd. This was, for Descartes, an irrefutable principle upon which to ground all forms of other knowledge. Descartes posited a metaphysical dualism, distinguishing between the substances of the human body ("*res extensa*") and the mind or soul ("*res cogitans*"). This crucial distinction would be left unresolved and lead to what is known as the mind-body problem, since the two substances in the Cartesian system are independent of each other and irreducible.

Baruch Spinoza (1632–1677)

The philosophy of Baruch Spinoza is a systematic, logical, rational philosophy developed in seventeenth-century Europe. Spinoza's philosophy is a system of ideas constructed upon basic building blocks with an internal consistency with which he tried to answer life's major questions and in which he proposed that "God exists only philosophically." He was heavily influenced by Descartes, Euclid and Thomas Hobbes, as well as theologians in the Jewish philosophical tradition such as Maimonides. But his work was in many respects a departure from the Judeo-Christian tradition. Many of Spinoza's ideas continue to vex thinkers today and many of his principles, particularly regarding the emotions, have implications for modern approaches to psychology. To this day, many important thinkers have found Spinoza's "geometrical method" difficult to comprehend: Goethe admitted that he found this concept confusing. His *magnum opus*, *Ethics*, contains unresolved obscurities and has a forbidding mathematical structure modeled on Euclid's geometry. Spinoza's philosophy attracted believers such as Albert Einstein and much intellectual attention.

Gottfried Leibniz (1646–1716)

Leibniz was the last of the great Rationalists who contributed heavily to other fields such as metaphysics, epistemology, logic, mathematics, physics, jurisprudence, and the philosophy of religion; he is also considered to be one of the last "universal geniuses". He did not develop his system, however, independently of these advances. Leibniz rejected Cartesian dualism and denied the existence of a material world. In Leibniz's view there are infinitely many simple substances, which he called "monads" (possibly taking the term from the work of Anne Conway).

Leibniz developed his theory of monads in response to both Descartes and Spinoza, because the rejection of their visions forced him to arrive at his own solution. Monads are the fundamental unit of reality, according to Leibniz, constituting both inanimate and animate objects. These units of reality represent the universe, though they are not subject to the laws of causality or space (which he called "well-founded phenomena"). Leibniz, therefore, introduced his principle of pre-established harmony to account for apparent causality in the world.

Immanuel Kant (1724–1804)

Kant is one of the central figures of modern philosophy, and set the terms by which all subsequent thinkers have had to grapple. He argued that human perception structures natural laws, and that reason is the source of morality. His thought continues to hold a major influence in contemporary thought, especially in fields such as metaphysics, epistemology, ethics, political philosophy, and aesthetics.

Kant named his brand of epistemology "Transcendental Idealism", and he first laid out these views in his famous work *The Critique of Pure Reason*. In it he argued that there were fundamental problems with both rationalist and empiricist dogma. To the rationalists he argued, broadly, that pure reason is flawed when it goes beyond its limits and claims to know those things that are necessarily beyond the realm of all possible experience: the existence of God, free will, and the immortality of the human soul. Kant referred to these objects as "The Thing in Itself" and goes on to argue that their status as objects beyond all possible experience by definition means we cannot know them. To the empiricist he argued that while it is correct that experience is fundamentally necessary for human knowledge, reason is necessary for processing that experience into coherent thought. He therefore concludes that both reason and experience are necessary for human knowledge. In the same way, Kant also argued that it was wrong to regard thought as mere analysis. In Kant's views, a priori concepts do exist, but if they are to lead to the amplification of knowledge, they must be brought into relation with empirical data".

Contemporary Rationalism

Rationalism has become a rarer label *tout court* of philosophers today; rather many different kinds of specialised rationalisms are identified. For example, Robert Brandom has appropriated the terms rationalist expressivism and rationalist pragmatism as labels for aspects of his programme in *Articulating Reasons*, and identified linguistic rationalism, the claim that the content of propositions "are essentially what can serve as both premises and conclusions of inferences", as a key thesis of Wilfred Sellars.

References

- Katsicas, Sokratis K. (2009). "Chapter 35". In Vacca, John. Computer and Information Security Handbook. Morgan Kaufmann Publications. Elsevier Inc. p. 605. ISBN 978-0-12-374354-1.
- Cottingham, J., ed. (April 1996) [1986]. Meditations on First Philosophy With Selections from the Objections and Replies (revised ed.). Cambridge University Press. ISBN 978-0-521-55818-1.
- Stanford Encyclopedia of Philosophy, Rationalism vs. Empiricism First published August 19, 2004; substantive revision March 31, 2013 cited on May 20, 2013.
- Boyd, Richard, "The Value of Civility?," Urban Studies Journal, May 2006, vol. 43 (no. 5–6), pp. 863–78 Retrieved 2013-01-13.
- Stanford Encyclopedia of Philosophy, The Intuition/Deduction Thesis First published August 19, 2004; substantive revision March 31, 2013 cited on May 20, 2013.
- Watson, Richard A. (31 March 2012). "René Descartes". Encyclopædia Britannica. Encyclopædia Britannica Online. Encyclopædia Britannica Inc. Retrieved 31 March 2012.
- "Immanuel Kant (Stanford Encyclopedia of Philosophy)". Plato.stanford.edu. 20 May 2010. Retrieved 2011-10-22.

9

Key Concepts in Methodology

The key concepts in methodology are research question, thesis statement and paradigm. The research question is the question that is asked in the research paper, which the research paper answers as its conclusion. In order to completely understand methodology, it is necessary to understand the processes related to it. The following chapter elucidates the main concepts in methodology.

Research Question

Specifying the research question is the methodological point of departure of scholarly research in both the natural and social sciences. The research will answer the question posed. At an undergraduate level, the answer to the research question is the thesis statement. The answer to a research question will help address a "Research Problem" which is a problem "readers think is worth solving".

Overview

Specifying the research question is one of the first methodological steps the investigator has to take when undertaking research. The research question must be accurately and clearly defined.

Choosing a research question is the central element of both quantitative and qualitative research and in some cases it may precede construction of the conceptual framework of study. In all cases, it makes the theoretical assumptions in the framework more explicit, most of all it indicates what the researcher wants to know most and first.

The student or researcher then carries out the research necessary to answer the research question, whether this involves reading secondary sources over a few days for an undergraduate term paper or carrying out primary research over years for a major project.

When the research is complete and the researcher knows the (probable) answer to the research question, writing up can begin (as distinct from writing notes, which is a process that goes on through a research project). In term papers, the answer to the question is normally given in summary in the introduction in the form of a thesis statement.

Types and Purpose

The research question serves two purposes:

1. It determines where and what kind of research the writer will be looking for.
2. It identifies the specific objectives the study or paper will address.

Therefore, the writer must first identify the type of study (qualitative, quantitative, or mixed) before the research question is developed.

Qualitative Study

A qualitative study seeks to learn why or how, so the writer's research must be directed at determining the what, why and how of the research topic. Therefore, when crafting a research question for a qualitative study, the writer will need to ask a why or how question about the topic. For example: How did the company successfully market its new product? The sources needed for qualitative research typically include print and internet texts (written words), audio and visual media.

Here is Creswell's (2009) example of a script for a qualitative research central question:

- __________ (How or what) is the __________ ("story for" for narrative research; "meaning of" the phenomenon for phenomenology; "theory that explains the process of" for grounded theory; "culture-sharing pattern" for ethnography; "issue" in the "case" for case study) of __________ (central phenomenon) for __________ (participants) at __________ (research site).

Quantitative Study

A quantitative study seeks to learn where, or when, so the writer's research must be directed at determining the where, or when of the research topic. Therefore, when crafting a research question for a quantitative study, the writer will need to ask a where, or when question about the topic. For example: Where should the company market its new product? Unlike a qualitative study, a quantitative study is mathematical analysis of the research topic, so the writer's research will consist of numbers and statistics.

Here is Creswell's (2009) example of a script for a quantitative research question:

- Does __________ (name the theory) explain the relationship between __________ (independent variable) and __________ (dependent variable), controlling for the effects of __________ (control variable)?

Alternatively, a script for a quantitative null hypothesis might be as follows:

- There is no significant difference between __________ (the control and experimental groups on the independent variable) on __________ (dependent variable).

Quantitative studies also fall into two categories:

1. Correlational studies: A correlational study is non-experimental, requiring the writer to research relationships without manipulating or randomly selecting the subjects of the research. The research question for a correlational study may look like this: What is the relationship between long distance commuters and eating disorders?
2. Experimental studies: An experimental study is experimental in that it requires the writer to manipulate and randomly select the subjects of the research. The research question for an experimental study may look like this: Does the consumption of fast food lead to eating disorders?

Mixed Study

A mixed study integrates both qualitative and quantitative studies, so the writer's research must be directed at determining the why or how and the what, where, or when of the research topic. Therefore, the writer will need to craft a research question for each study required for the assignment. Note: A typical study may be expected to have between 1 and 6 research questions.

Once the writer has determined the type of study to be used and the specific objectives the paper will address, the writer must also consider whether the research question passes the 'so what' test. The 'so what' test means that the writer must construct evidence to convince the audience why the research is expected to add new or useful knowledge to the literature.

Related Terms

Problematique

Problematique is a term that functions analogously to the research problem or question used typically when addressing global systemic problems. The term achieved prominence in 1970 when Hasan Özbekhan, Erich Jantsch and Alexander Christakis conceptualized the original prospectus of the Club of Rome titled "The Predicament of Mankind". In this prospectus the authors designated 49 Continuous Critical Problems facing humankind, saying "We find it virtually impossible to view them as problems that exist in isolation - or as problems capable of being solved in their own terms... It is this generalized meta system of problems, which we call the 'problematique' that inheres in our situation."

Situations similar to the global problematique in their complexity are also called problematiques. These situations receive different designations from other authors. C. West Churchman, Rittell and Weber, and Argyris call these situations wicked problems. Russell Ackoff simply called them "messes.'"

Thesis Statement

A thesis statement usually appears at the end of the introductory paragraph of a paper, and it offers a concise summary of the main point or claim of the essay, research paper, etc. A thesis statement is usually one sentence that appears at the beginning, though it may occur more than once. The thesis statement is developed, supported, and explained in the course of the paper by means of examples and evidence. Thesis statements help organize and develop the system of proper writing, and also serve as a signal to readers about the topic of a paper. They are essential components of scholarly research papers.

Types of Thesis Statements

The thesis statement will reflect the kind of paper being written. There are 3 kinds of papers: analytical, expository, and argumentative. The thesis statement will take a different form for each of these kinds of papers.

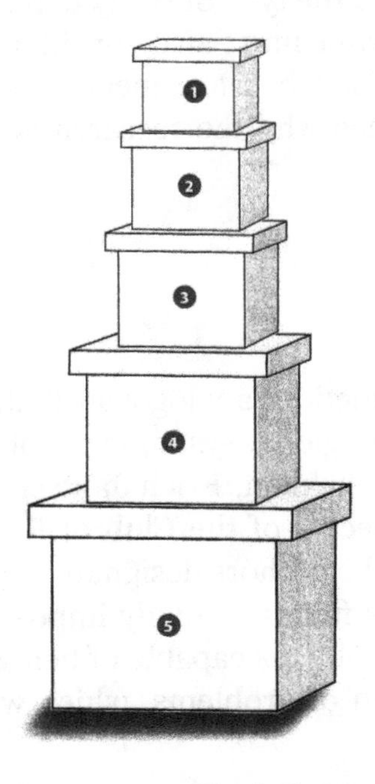

Narrow Down to a Thesis Statement

Paradigm

In science and philosophy, is a distinct set of concepts or thought patterns, including theories, research methods, postulates, and standards for what constitutes legitimate contributions to a field.

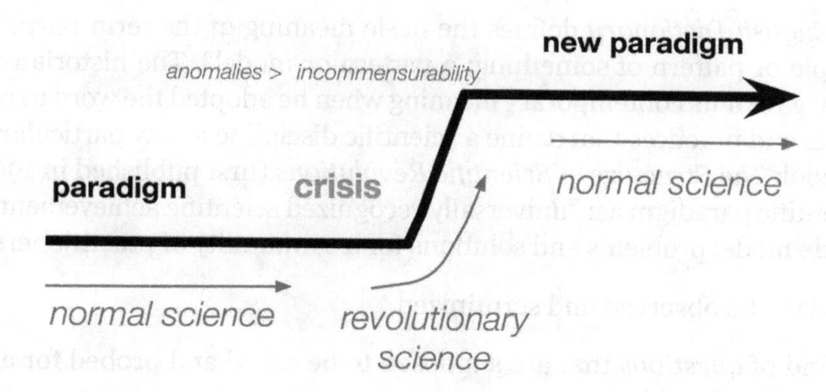

Etymology

In rhetoric, paradeigma is known as a type of proof. The purpose of paradeigma is to provide an audience with an illustration of similar occurrences. This illustration is not meant to take the audience to a conclusion, however it is used to help guide them there. A personal accountant is a good comparison of paradeigma to explain how it is meant to guide the audience. It is not the job of a personal accountant to tell their client exactly what (and what not) to spend their money on, but to aid in guiding their client as to how money should be spent based on their financial goals. Anaximenes defined paradeigma as, "actions that have occurred previously and are similar to, or the opposite of, those which we are now discussing."

The term had a technical meaning in the field of grammar: the 1900 *Merriam-Webster* dictionary defines its technical use only in the context of grammar or, in rhetoric, as a term for an illustrative parable or fable. In linguistics, Ferdinand de Saussure used *paradigm* to refer to a class of elements with similarities.

The Merriam-Webster Online dictionary defines this usage as "a philosophical and theoretical framework of a scientific school or discipline within which theories, laws, and generalizations and the experiments performed in support of them are formulated; *broadly: a philosophical or theoretical framework of any kind.*"

The Oxford Dictionary of Philosophy attributes the following description of the term to Thomas Kuhn's The Structure of Scientific Revolutions:

Kuhn suggests that certain scientific works, such as Newton's Principia or John Dalton's New System of Chemical Philosophy (1808), provide an open-ended resource: a framework of concepts, results, and procedures within which subsequent work is structured. Normal science proceeds within such a framework or paradigm. A paradigm does not impose a rigid or mechanical approach, but can be taken more or less creatively and flexibly.

Scientific Paradigm

The *Oxford English Dictionary* defines the basic meaning of the term *paradigm* as "a typical example or pattern of something; a pattern or model". The historian of science Thomas Kuhn gave it its contemporary meaning when he adopted the word to refer to the set of concepts and practices that define a scientific discipline at any particular period of time. In his book *The Structure of Scientific Revolutions* (first published in 1962), Kuhn defines a scientific paradigm as: "universally recognized scientific achievements that, for a time, provide model problems and solutions for a community of practitioners, i.e.,

- *what* is to be observed and scrutinized
- the kind of *questions* that are supposed to be asked and probed for answers in relation to this subject
- *how* these questions are to be structured
- *what* predictions made by the primary theory within the discipline
- *how* the results of scientific investigations should be interpreted
- *how* an experiment is to be conducted, and *what* equipment is available to conduct the experiment.

In *The Structure of Scientific Revolutions*, Kuhn saw the sciences as going through alternating periods of *normal science*, when an existing model of reality dominates a protracted period of puzzle-solving, and *revolution*, when the model of reality itself undergoes sudden drastic change. Paradigms have two aspects. Firstly, within normal science, the term refers to the set of exemplary experiments that are likely to be copied or emulated. Secondly, underpinning this set of exemplars are shared preconceptions, made prior to – and conditioning – the collection of evidence. These preconceptions embody both hidden assumptions and elements that he describes as quasi-metaphysical; the interpretations of the paradigm may vary among individual scientists.

Kuhn was at pains to point out that the rationale for the choice of exemplars is a specific way of viewing reality: that view and the status of "exemplar" are mutually reinforcing. For well-integrated members of a particular discipline, its paradigm is so convincing that it normally renders even the possibility of alternatives unconvincing and counter-intuitive. Such a paradigm is *opaque*, appearing to be a direct view of the bedrock of reality itself, and obscuring the possibility that there might be other, alternative imageries hidden behind it. The conviction that the current paradigm *is* reality tends to disqualify evidence that might undermine the paradigm itself; this in turn leads to a build-up of unreconciled anomalies. It is the latter that is responsible for the eventual revolutionary overthrow of the incumbent paradigm, and its replacement by a new one. Kuhn used the expression *paradigm shift* for this process, and likened it to the perceptual change that occurs when our interpretation of an ambiguous image "flips over" from one state to another. (The rabbit-duck illusion is an example: it is not possible to see both the rabbit and the duck simultaneously.) This is significant in relation to the issue of *incommensurability*.

An example of a currently accepted paradigm would be the standard model of physics. The scientific method allows for orthodox scientific investigations into phenomena that might contradict or disprove the standard model; however grant funding would be proportionately more difficult to obtain for such experiments, depending on the degree of deviation from the accepted standard model theory the experiment would test for. To illustrate the point, an experiment to test for the mass of neutrinos or the decay of protons (small departures from the model) is more likely to receive money than experiments that look for the violation of the conservation of momentum, or ways to engineer reverse time travel.

Mechanisms similar to the original Kuhnian paradigm have been invoked in various disciplines other than the philosophy of science. These include: the idea of major cultural themes, worldviews, ideologies, and mindsets. They have somewhat similar meanings that apply to smaller and larger scale examples of disciplined thought. In addition, Michel Foucault used the terms episteme and discourse, mathesis and taxinomia, for aspects of a "paradigm" in Kuhn's original sense.

Paradigm Shifts

In *The Structure of Scientific Revolutions*, Kuhn wrote that "the successive transition from one paradigm to another via revolution is the usual developmental pattern of mature science."

Paradigm shifts tend to appear in response to the accumulation of critical anomalies as well as the proposal of a new theory with the power to encompass both older relevant data and explain relevant anomalies. New paradigms tend to be most dramatic in sciences that appear to be stable and mature, as in physics at the end of the 19th century. At that time, a statement generally attributed to physicist Lord Kelvin famously claimed, "There is nothing new to be discovered in physics now. All that remains is more and more precise measurement." Five years later, Albert Einstein published his paper on special relativity, which challenged the set of rules laid down by Newtonian mechanics, which had been used to describe force and motion for over two hundred years. In this case, the new paradigm reduces the old to a special case in the sense that Newtonian mechanics is still a good model for approximation for speeds that are slow compared to the speed of light. Many philosophers and historians of science, including Kuhn himself, ultimately accepted a modified version of Kuhn's model, which synthesizes his original view with the gradualist model that preceded it. Kuhn's original model is now generally seen as too limited.

Kuhn's idea was itself revolutionary in its time, as it caused a major change in the way that academics talk about science. Thus, it may be that it caused or was itself part of a "paradigm shift" in the history and sociology of science. However, Kuhn would not recognize such a paradigm shift. Being in the social sciences, people can still use earlier ideas to discuss the history of science.

Paradigm Paralysis

Perhaps the greatest barrier to a paradigm shift, in some cases, is the reality of paradigm paralysis: the inability or refusal to see beyond the current models of thinking. This is similar to what psychologists term Confirmation bias. Examples include rejection of Aristarchus of Samos', Copernicus', and Galileo's theory of a heliocentric solar system, the discovery of electrostatic photography, xerography and the quartz clock.

Incommensurability

Kuhn pointed out that it could be difficult to assess whether a particular paradigm shift had actually led to progress, in the sense of explaining more facts, explaining more important facts, or providing better explanations, because the understanding of "more important", "better", etc. changed with the paradigm. The two versions of reality are thus *incommensurable*. Kuhn's version of incommensurability has an important psychological dimension; this is apparent from his analogy between a paradigm shift and the flip-over involved in some optical illusions. However, he subsequently diluted his commitment to incommensurability considerably, partly in the light of other studies of scientific development that did not involve revolutionary change. One of the examples of incommensurability that Kuhn used was the change in the style of chemical investigations that followed the work of Lavoisier on atomic theory in the late 18th Century. In this change, the focus had shifted from the bulk properties of matter (such as hardness, colour, reactivity, etc.) to studies of atomic weights and quantitative studies of reactions. He suggested that it was impossible to make the comparison needed to judge which body of knowledge was better or more advanced. However, this change in research style (and paradigm) eventually (after more than a century) led to a theory of atomic structure that accounts well for the bulk properties of matter; see, for example, Brady's *General Chemistry*. According to P J Smith, this ability of science to back off, move sideways, and then advance is characteristic of the natural sciences, but contrasts with the position in some social sciences, notably economics.

This apparent ability does not guarantee that the account is veridical at any one time, of course, and most modern philosophers of science are fallibilists. However, members of other disciplines do see the issue of incommensurability as a much greater obstacle to evaluations of "progress"; see, for example, Martin Slattery's *Key Ideas in Sociology*.

Subsequent Developments

Opaque Kuhnian paradigms and paradigm shifts do exist. A few years after the discovery of the mirror-neurons that provide a hard-wired basis for the human capacity for empathy, the scientists involved were unable to identify the incidents that had directed their attention to the issue. Over the course of the investigation, their language and metaphors had changed so that they themselves could no longer interpret all of their own earlier laboratory notes and records.

Imre Lakatos and Research Programmes

However, many instances exist in which change in a discipline's core model of reality has happened in a more evolutionary manner, with individual scientists exploring the usefulness of alternatives in a way that would not be possible if they were constrained by a paradigm. Imre Lakatos suggested (as an alternative to Kuhn's formulation) that scientists actually work within research programmes. In Lakatos' sense, a research programme is a sequence of problems, placed in order of priority. This set of priorities, and the associated set of preferred techniques, is the positive heuristic of a programme. Each programme also has a negative heuristic; this consists of a set of fundamental assumptions that – temporarily, at least – takes priority over observational evidence when the two appear to conflict.

This latter aspect of research programmes is inherited from Kuhn's work on paradigms, and represents an important departure from the elementary account of how science works. According to this, science proceeds through repeated cycles of observation, induction, hypothesis-testing, etc., with the test of consistency with empirical evidence being imposed at each stage. Paradigms and research programmes allow anomalies to be set aside, where there is reason to believe that they arise from incomplete knowledge (about either the substantive topic, or some aspect of the theories implicitly used in making observations.

Larry Laudan: Dormant Anomalies, Fading Credibility, and Research Traditions

Larry Laudan has also made two important contributions to the debate. Laudan believed that something akin to paradigms exist in the social sciences; he referred to these as research traditions. Laudan noted that some anomalies become "dormant", if they survive a long period during which no competing alternative has shown itself capable of resolving the anomaly. He also presented cases in which a dominant paradigm had withered away because its lost credibility when viewed against changes in the wider intellectual milieu.

In Social Sciences

Kuhn himself did not consider the concept of paradigm as appropriate for the social sciences. He explains in his preface to *The Structure of Scientific Revolutions* that he developed the concept of paradigm precisely to distinguish the social from the natural sciences. While visiting the Center for Advanced Study in the Behavioral Sciences in 1958 and 1959, surrounded by social scientists, he observed that they were never in agreement about the nature of legitimate scientific problems and methods. He explains that he wrote this book precisely to show that there can never be any paradigms in the social sciences. Mattei Dogan, a French sociologist, in his article "Paradigms in the Social Sciences," develops Kuhn's original thesis that there are no paradigms at all in the social sciences since the concepts are polysemic, involving the deliberate mutual ignorance between scholars and the proliferation of schools in these disciplines. Dogan provides many examples of the non-existence of paradigms in the social sciences in his essay, particularly in sociology, political science and political anthropology.

However, both Kuhn's original work and Dogan's commentary are directed at disciplines that are defined by conventional labels (such as well as "sociology"). While it is true that such broad groupings in the social sciences are usually not based on a Kuhnian paradigm, each of the competing sub-disciplines may still be underpinned by a paradigm, research programme, research tradition, and/ or professional imagery. These structures will be motivating research, providing it with an agenda, defining what is and is not anomalous evidence, and inhibiting debate with other groups that fall under the same broad disciplinary label. (A good example is provided by the contrast between Skinnerian behaviourism and Personal Construct Theory (PCT) within psychology. The most significant of the many ways these two sub-disciplines of psychology differ concerns meanings and intentions. In PCT, they are seen as the central concern of psychology; in behaviourism, they are not scientific evidence at all, as they cannot be directly observed.)

Such considerations explain the conflict between the Kuhn/ Dogan view, and the views of others (including Larry Laudan), who do apply these concepts to social sciences.

Handa, M.L. (1986) introduced the idea of "social paradigm" in the context of social sciences. He identified the basic components of a social paradigm. Like Kuhn, Handa addressed the issue of changing paradigm; the process popularly known as "paradigm shift". In this respect, he focused on social circumstances that precipitate such a shift and the effects of the shift on social institutions, including the institution of education. This broad shift in the social arena, in turn, changes the way the individual perceives reality.

Another use of the word *paradigm* is in the sense of "worldview". For example, in social science, the term is used to describe the set of experiences, beliefs and values that affect the way an individual perceives reality and responds to that perception. Social scientists have adopted the Kuhnian phrase "paradigm shift" to denote a change in how a given society goes about organizing and understanding reality. A "dominant paradigm" refers to the values, or system of thought, in a society that are most standard and widely held at a given time. Dominant paradigms are shaped both by the community's cultural background and by the context of the historical moment. The following are conditions that facilitate a system of thought to become an accepted dominant paradigm:

- Professional organizations that give legitimacy to the paradigm
- Dynamic leaders who introduce and purport the paradigm
- Journals and editors who write about the system of thought. They both disseminate the information essential to the paradigm and give the paradigm legitimacy
- Government agencies who give credence to the paradigm
- Educators who propagate the paradigm's ideas by teaching it to students
- Conferences conducted that are devoted to discussing ideas central to the paradigm
- Media coverage
- Lay groups, or groups based around the concerns of lay persons, that embrace the beliefs central to the paradigm
- Sources of funding to further research on the paradigm

Other Uses

The word *paradigm* is also still used to indicate a pattern or model or an outstandingly clear or typical example or archetype. The term is frequently used in this sense in the design professions. Design Paradigms or archetypes comprise functional precedents for design solutions. The best known references on design paradigms are *Design Paradigms: A Sourcebook for Creative Visualization*, by Wake, and *Design Paradigms* by Petroski.

This term is also used in cybernetics. Here it means (in a very wide sense) a (conceptual) protoprogram for reducing the chaotic mass to some form of order. Note the similarities to the concept of entropy in chemistry and physics. A paradigm there would be a sort of prohibition to proceed with any action that would increase the total entropy of the system. To create a paradigm requires a closed system that accepts changes. Thus a paradigm can only apply to a system that is not in its final stage.

Beyond its use in the physical and social sciences, Kuhn's paradigm concept has been analysed in relation to its applicability in identifying 'paradigms' with respect to worldviews at specific points in history. One example is Matthew Edward Harris' book *The*

Notion of Papal Monarchy in the Thirteenth Century: The Idea of Paradigm in Church History. Harris stresses the primarily sociological importance of paradigms, pointing towards Kuhn's second edition of *The Structure of Scientific Revolutions*. Although obedience to popes such as Innocent III and Boniface VIII was widespread, even written testimony from the time showing loyalty to the pope does not demonstrate that the writer had the same worldview as the Church, and therefore pope, at the centre. The difference between paradigms in the physical sciences and in historical organisations such as the Church is that the former, unlike the latter, requires technical expertise rather than repeating statements. In other words, after scientific training through what Kuhn calls 'exemplars', one could not genuinely believe that, to take a trivial example, the earth is flat, whereas thinkers such as Giles of Rome in the thirteenth century wrote in favour of the pope, then could easily write similarly glowing things about the king. A writer such as Giles would have wanted a good job from the pope; he was a papal publicist. However, Harris writes that 'scientific group membership is not concerned with desire, emotions, gain, loss and any idealistic notions concerning the nature and destiny of humankind...but simply to do with aptitude, explanation, [and] cold description of the facts of the world and the universe from within a paradigm'.

References

- Booth, Wayne (1995). The Craft of Research. Chicago, IL: The University of Chicago Press. ISBN 0226065650.
- Sampley, J. Paul (2003). Paul in the Greco-Roman World: A Handbook. Trinity Press International. pp. 228–229. ISBN 9781563382666.
- Blackburn, Simon, 1994, 2005, 2008, rev. 2nd ed. The Oxford Dictionary of Philosophy. Oxford: Oxford University Press. ISBN 0-19-283134-8. Description & 1994 letter-preview links.
- Kuhn, T S (1970) The Structure of Scientific Revolutions (2nd Edition) University of Chicago Press. Section V, pages 43-51. ISBN 0-226-45804-0.
- Haack, S (2003) Defending Science – within reason: between scientism and cynicism. Prometheus Books. ISBN 978-1-59102-458-3.
- Slattery, Martin (2003). Key ideas in sociology. OCLC Number: 52531237. Cheltenham : Nelson Thornes. pp. 151, 152, 153, 155. ISBN 978-0-7487-6565-2.
- Harris, Matthew (2010). The notion of papal monarchy in the thirteenth century : the idea of paradigm in church history. Lewiston, N.Y.: Edwin Mellen Press. p. 160. ISBN 978-0-7734-1441-9.
- "Tips and Examples for Writing Thesis Statements". Purdue OWL. Purdue University. Retrieved 6 October 2014

10

Philosophical Methodology: A Comprehensive Study

Philosophical methodology is the method of doing philosophy. Some of the methods that philosophers follow are methodic doubt, argument and dialectic. The themes discussed are phenomenology, quietism, methodism, experimental philosophy and intercultural philosophy. These topics are crucial for a complete understanding of the subject of research design and methodology.

Philosophical Methodology

Philosophical method (or philosophical methodology) is the study of how to do philosophy. A common view among philosophers is that philosophy is distinguished by the ways that philosophers follow in addressing philosophical questions. There is not just one method that philosophers use to answer philosophical questions.

Methodology Process

Systematic philosophy is a generic term that applies to philosophical methods and approaches that attempt to provide a framework in reason that can explain all questions and problems related to human life. Examples of systematic philosophers include Plato, Aristotle, Descartes, Spinoza, and Hegel. In many ways, any attempts to formulate a philosophical method that provides the ultimate constituents of reality, a metaphysics, can be considered systematic philosophy. In modern philosophy the reaction to systematic philosophy began with Kierkegaard and continued in various forms through analytic philosophy, existentialism, hermeneutics, and deconstructionism.

Some common features of the methods that philosophers follow (and discuss when discussing philosophical method) include:

- Methodic doubt - a systematic process of being skeptical about (or doubting) the truth of one's beliefs.
- Argument - provide an argument or several arguments supporting the solution.
- Dialectic - present the solution and arguments for criticism by other philosophers, and help them judge their own.

Doubt and the Sense of Wonder

Plato said that "philosophy begins in wonder", a view which is echoed by Aristotle: "It was their wonder, astonishment, that first led men to philosophize and still leads them." Philosophizing may begin with some simple doubts about accepted beliefs. The initial impulse to philosophize may arise from suspicion, for example that we do not fully understand, and have not fully justified, even our most basic beliefs about the world.

Formulate Questions and Problems

Another element of philosophical method is to formulate questions to be answered or problems to be solved. The working assumption is that the more clearly the question or problem is stated, the easier it is to identify critical issues.

A relatively small number of major philosophers prefer not to be quick, but to spend more time trying to get extremely clear on what the problem is all about.

Enunciate a Solution

Another approach is to enunciate a theory, or to offer a definition or analysis, which constitutes an attempt to solve a philosophical problem. Sometimes a philosophical theory by itself can be stated quite briefly. All the supporting philosophical text is offered by way of hedging, explanation, and argument.

Not all proposed solutions to philosophical problems consist of definitions or generalizations. Sometimes what is called for is a certain sort of explanation — not a causal explanation, but an explanation for example of how two different views, which seem to be contrary to one another, can be held at the same time, consistently. One can call this a philosophical explanation.

Justify the Solution

An argument is a set of statements, one of which (the conclusion), it is said or implied, follows from the others (the premises). One might think of arguments as bundles of reasons — often not just a list, but logically interconnected statements — followed by the claim they are reasons for. The reasons are the premises, the claim they support is the conclusion; together they make an argument.

Philosophical arguments and justifications are another important part of philosophical method. It is rare to find a philosopher, particularly in the Western philosophical tradition, who lacks many arguments. Philosophers are, or at least are expected to be, very good at giving arguments. They constantly demand and offer arguments for different claims they make. This therefore indicates that philosophy is a quest for arguments.

A good argument — a clear, organized, and sound statement of reasons — may ultimately cure the original doubts that motivated us to take up philosophy. If one is willing to be satisfied without any good supporting reasons, then a Western philosophical approach may not be what one actually requires.

Philosophical Criticism

In philosophy, which concerns the most fundamental aspects of the universe, the experts all disagree. It follows that another element of philosophical method, common in the work of nearly all philosophers, is philosophical criticism. It is this that makes much philosophizing a social endeavor.

Philosophers offer definitions and explanations in solution to problems; they argue for those solutions; and then other philosophers provide counter arguments, expecting to eventually come up with better solutions. This exchange and resulting revision of views is called dialectic. Dialectic (in one sense of this history-laden word) is simply philosophical conversation amongst people who do not always agree with each other about everything.

One can do this sort of harsh criticism on one's own, but others can help greatly, if important assumptions are shared with the person offering the criticisms. Others are able to think of criticisms from another perspective.

Some philosophers and ordinary people dive right in and start trying to solve the problem. They immediately start giving arguments, pro and con, on different sides of the issue. Doing philosophy is different from this. It is about questioning assumptions, digging for deeper understanding. Doing philosophy is about the journey, the process, as much as it is about the destination, the conclusion. Its method differs from other disciplines, in which the experts can agree about most of the fundamentals.

Motivation

Method in philosophy is in some sense rooted in motivation, only by understanding why people take up philosophy can one properly understand what philosophy is. People often find themselves believing things that they do not understand. For example, about God, themselves, the natural world, human society, morality and human productions. Often, people fail to understand what it is they believe, and fail to understand the reasons they believe in what they do. Some people have questions about the meaning of their beliefs and questions about the justification (or rationality) of their beliefs. A lack of these things shows a lack of understanding, and some dislike not having this understanding.

These questions about are only the tip of the philosophical iceberg. There are many other things about this universe about which people are also fundamentally ignorant. Philosophers are in the business of investigating all sorts of those areas of ignorance.

A bewilderingly huge number of basic concepts are poorly understood. For example:

- What does it mean to say that one thing causes another?
- What is rationality? What are space and time?
- What is beauty, and if it is in the eye of the beholder, then what is it that is being said to be in the eye of the beholder?

One might also consider some of the many questions about justification. Human lives are deeply informed with many basic assumptions. Different assumptions would lead to different ways of living.

Phenomenology (Philosophy)

Phenomenology (from Greek *phainómenon* "that which appears" and *lógos* "study") is the philosophical study of the structures of experience and consciousness. As a philosophical movement it was founded in the early years of the 20th century by Edmund Husserl and was later expanded upon by a circle of his followers at the universities of Göttingen and Munich in Germany. It then spread to France, the United States, and elsewhere, often in contexts far removed from Husserl's early work. Phenomenology should not be considered as a unitary movement; rather, different authors share a common family resemblance but also with many significant differences. Accordingly, "A unique and final definition of phenomenology is dangerous and perhaps even paradoxical as it lacks a thematic focus. In fact, it is not a doctrine, nor a philosophical school, but rather a style of thought, a method, an open and ever-renewed experience having different results, and this may disorient anyone wishing to define the meaning of phenomenology".

Phenomenology, in Husserl's conception, is primarily concerned with the systematic reflection on and study of the structures of consciousness and the phenomena that appear in acts of consciousness. Phenomenology can be clearly differentiated from the Cartesian method of analysis which sees the world as objects, sets of objects, and objects acting and reacting upon one another.

Husserl's conception of phenomenology has been criticized and developed not only by himself but also by students, such as Edith Stein and Roman Ingarden, by hermeneutic philosophers, such as Martin Heidegger, by existentialists, such as Nicolai Hartmann, Gabriel Marcel, Maurice Merleau-Ponty, Jean-Paul Sartre, and by other philosophers, such as Max Scheler, Paul Ricoeur, Jean-Luc Marion, Emmanuel Levinas, and sociologists Alfred Schütz and Eric Voegelin.

Overview

In its most basic form, phenomenology attempts to create conditions for the objective study of topics usually regarded as subjective: consciousness and the content of conscious experiences such as judgments, perceptions, and emotions. Although phenomenology seeks to be scientific, it does not attempt to study consciousness from the perspective of clinical psychology or neurology. Instead, it seeks through systematic reflection to determine the essential properties and structures of experience.

There are several assumptions behind phenomenology that help explain its foundations:

1. It rejects the concept of objective research. Phenomenologists prefer grouping assumptions through a process called phenomenological epoche.
2. Phenomenology believes that analyzing daily human behavior can provide one with a greater understanding of nature.
3. Persons should be explored. This is because persons can be understood through the unique ways they reflect the society they live in.
4. Phenomenologists prefer to gather "capta," or conscious experience, rather than traditional data.
5. Phenomenology is considered to be oriented on discovery, and therefore phenomenologists gather research using methods that are far less restricting than in other sciences.

Husserl derived many important concepts central to phenomenology from the works and lectures of his teachers, the philosophers and psychologists Franz Brentano and Carl Stumpf. An important element of phenomenology that Husserl borrowed from Brentano is intentionality (often described as "aboutness"), the notion that consciousness is always consciousness *of* something. The object of consciousness is called the *intentional object*, and this object is constituted for consciousness in many different ways, through, for instance, perception, memory, retention and protention, signification, etc. Throughout these different intentionalities, though they have different structures and different ways of being "about" the object, an object is still constituted as the identical object; consciousness is directed at the same intentional object in direct perception as it is in the immediately following retention of this object and the eventual remembering of it.

Though many of the phenomenological methods involve various reductions, phenomenology is, in essence, anti-reductionistic; the reductions are mere tools to better understand and describe the workings of consciousness, not to reduce any phenomenon to these descriptions. In other words, when a reference is made to a thing's *essence* or *idea*, or when one details the constitution of an identical coherent thing by describing what one "really" sees as being only these sides and aspects, these surfaces, it does not mean that the thing is only and exclusively what is described here: The ultimate goal of these reductions is to understand *how* these different aspects are constituted into the actual thing as experienced by the person experiencing it. Phenomenology is a direct reaction to the psychologism and physicalism of Husserl's time.

Although previously employed by Georg Wilhelm Friedrich Hegel in his *Phenomenology of Spirit*, it was Husserl's adoption of this term (circa 1900) that propelled it into becoming the designation of a philosophical school. As a philosophical perspective, phenomenology is its method, though the specific meaning of the term varies according to how it is conceived by a given philosopher. As envisioned by Husserl, phenomenology is a method of philosophical inquiry that rejects the rationalist bias that has dominated Western thought since Plato in favor of a method of reflective attentiveness that discloses the individual's "lived experience." Loosely rooted in an epistemological device, with Sceptic roots, called epoché, Husserl's method entails the suspension of judgment while relying on the intuitive grasp of knowledge, free of presuppositions and intellectualizing. Sometimes depicted as the "science of experience," the phenomenological method is rooted in intentionality, Husserl's theory of consciousness (developed from Brentano). Intentionality represents an alternative to the representational theory of consciousness, which holds that reality cannot be grasped directly because it is available only through perceptions of reality that are representations of it in the mind. Husserl countered that consciousness is not "in" the mind but rather conscious of something other than itself (the intentional object), whether the object is a substance or a figment of imagination (i.e., the real processes associated with and underlying the figment). Hence the phenomenological method relies on the description of phenomena as they are given to consciousness, in their immediacy.

According to Maurice Natanson (1973, p. 63), "The radicality of the phenomenological method is both continuous and discontinuous with philosophy's general effort to subject experience to fundamental, critical scrutiny: to take nothing for granted and to show the warranty for what we claim to know." In practice, it entails an unusual combination of discipline and detachment to suspend, or bracket, theoretical explanations and second-hand information while determining one's "naive" experience of the matter. The phenomenological method serves to momentarily erase the world of speculation by returning the subject to his or her primordial experience of the matter, whether the object of inquiry is a feeling, an idea, or a perception. According to Husserl the suspension of belief in what we ordinarily take for granted or infer by conjecture diminishes the power of what we customarily embrace as objective reality. According to Rüdiger Safranski (1998, 72), "[Husserl and his followers'] great ambition was to disregard anything that had until then been thought or said about consciousness or the world [while] on the lookout for a new way of letting the things [they investigated] approach them, without covering them up with what they already knew."

Martin Heidegger modified Husserl's conception of phenomenology because of (what Heidegger perceived as) Husserl's subjectivist tendencies. Whereas Husserl conceived humans as having been constituted by states of consciousness, Heidegger countered that consciousness is peripheral to the primacy of one's existence (i.e., the mode of being of Dasein), which cannot be reduced to one's consciousness of it. From this angle, one's state of mind is an "effect" rather than a determinant of existence, including those aspects of existence that one is not conscious of. By shifting the center of gravity from consciousness (psychology) to existence (ontology), Heidegger altered the subsequent direction of phenomenology. As one consequence of Heidegger's modification of Husserl's conception, phenomenology became increasingly relevant to psychoanalysis. Whereas Husserl gave priority to a depiction of consciousness that was fundamentally alien to the psychoanalytic conception of the unconscious, Heidegger offered a way to conceptualize experience that could accommodate those aspects of one's existence that lie on the periphery of sentient awareness.

Historical Overview of the Use of the Term

Phenomenology has at least two main meanings in philosophical history: one in the writings of G. W. F. Hegel, another in the writings of Edmund Husserl in 1920, and thirdly, succeeding Husserl's work, in the writings of his former research assistant Martin Heidegger in 1927.

- For G. W. F. Hegel, phenomenology is an approach to philosophy that begins with an exploration of phenomena (what presents itself to us in conscious experience) as a means to finally grasp the absolute, logical, ontological and metaphysical Spirit that is behind phenomena. This has been called dialectical phenomenology.
- For Edmund Husserl, phenomenology is "the reflective study of the essence of consciousness as experienced from the first-person point of view." Phenomenology takes the intuitive experience of phenomena (what presents itself to us in phenomenological reflexion) as its starting point and tries to extract from it the essential features of experiences and the essence of what we experience. When generalized to the essential features of any possible experience, this has been called transcendental phenomenology. Husserl's view was based on aspects of the work of Franz Brentano and was developed further by philosophers such as Maurice Merleau-Ponty, Max Scheler, Edith Stein, Dietrich von Hildebrand and Emmanuel Levinas.

Although the term "phenomenology" was used occasionally in the history of philosophy before Husserl, modern use ties it more explicitly to his particular method. Following is a list of important thinkers in rough chronological order who used the term "phenomenology" in a variety of ways, with brief comments on their contributions:

- Friedrich Christoph Oetinger (1702–1782), German pietist, for the study of the "divine system of relations"
- Johann Heinrich Lambert (1728–1777), mathematician, physician and philosopher, known for the theory of appearances underlying empirical knowledge.

- Immanuel Kant (1724–1804), in the Critique of Pure Reason, distinguished between objects as phenomena, which are objects as shaped and grasped by human sensibility and understanding, and objects as *things-in-themselves* or noumena, which do not appear to us in space and time and about which we can make no legitimate judgments.
- G. W. F. Hegel (1770–1831) challenged Kant's doctrine of the unknowable thing-in-itself, and declared that by knowing phenomena more fully we can gradually arrive at a consciousness of the absolute and spiritual truth of Divinity, most notably in his *Phenomenology of Spirit*, published in 1807.
- Carl Stumpf (1848–1936), student of Brentano and mentor to Husserl, used "phenomenology" to refer to an ontology of sensory contents.
- Edmund Husserl (1859–1938) established phenomenology at first as a kind of "descriptive psychology" and later as a transcendental and eidetic science of consciousness. He is considered to be the founder of contemporary phenomenology.
- Max Scheler (1874–1928) developed further the phenomenological method of Edmund Husserl and extended it to include also a reduction of the scientific method. He influenced the thinking of Pope John Paul II, Dietrich von Hildebrand, and Edith Stein.
- Martin Heidegger (1889–1976) criticized Husserl's theory of phenomenology and attempted to develop a theory of ontology that led him to his original theory of Dasein, the non-dualistic human being.
- Alfred Schütz (1899–1959) developed a phenomenology of the social world on the basis of everyday experience that has influenced major sociologists such as Harold Garfinkel, Peter Berger, and Thomas Luckmann.
- Francisco Varela (1946–2001), Chilean philosopher and biologist. Developed the basis for experimental phenomenology and neurophenomenology.

Later usage is mostly based on or (critically) related to Husserl's introduction and use of the term. This branch of philosophy differs from others in that it tends to be more "descriptive" than "prescriptive".

Varieties of Phenomenology

The *Encyclopedia of Phenomenology* (Kluwer Academic Publishers, 1997, Dordrecht and Boston) features separate articles on the following seven types of phenomenology: (1) Transcendental constitutive phenomenology studies how objects are constituted in transcendental consciousness, setting aside questions of any relation to the natural world. (2) Naturalistic constitutive phenomenology studies how consciousness constitutes things in the world of nature, assuming with the natural attitude that consciousness is part of nature. (3) Existential phenomenology studies concrete human existence, including our experience of free choice and/or action in concrete situations. (4) Generative historicist phenomenology studies how meaning—as found in our experience—is generated in historical processes of collective experience over time. (5) Genetic phenomenology studies the emergence/genesis of meanings of things within one's own stream of experience. (6) Hermeneutical phenomenology (also hermeneutic phenomenology or post-phenomenology/postphenomenology elsewhere) studies interpretive structures of experience. (7) Realistic phenomenology (also realist phenomenology elsewhere) studies the structure of consciousness and intentionality as "it occurs in a real world that is largely external to consciousness and not somehow brought into being by consciousness."

The contrast between "constitutive phenomenology" (German: *konstitutive Phänomenologie*; also static phenomenology (*statische Phänomenologie*)) or descriptive phenomenology (*beschreibende Phänomenologie*)) and "genetic phenomenology" (*genetische Phänomenologie*; also phenomenology of genesis (*Phänomenologie der Genesis*)) is due to Husserl.

Modern scholarship also recognizes the existence of the following varieties: late Heidegger's transcendental hermeneutic phenomenology, Maurice Merleau-Ponty's embodied phenomenology, and Michel Henry's material phenomenology.

Phenomenological Terminology

Intentionality

Intentionality refers to the notion that consciousness is always the consciousness *of* something. The word itself should not be confused with the "ordinary" use of the word intentional, but should rather be taken as playing on the etymological roots of the word. Originally, intention referred to a "stretching out" ("in tension," from Latin *intendere*), and in this context it refers to consciousness "stretching out" towards its object. However, one should be careful with this image: there is not some consciousness first that, subsequently, stretches out to its object; rather, consciousness *occurs as* the simultaneity of a conscious act and its object.

Intentionality is often summed up as "aboutness." Whether this *something* that consciousness is about is in direct perception or in fantasy is inconsequential to the concept of intentionality itself; whatever consciousness is directed at, *that* is what consciousness is conscious of. This means that the object of consciousness doesn't *have* to be a *physical* object apprehended in perception: it can just as well be a fantasy or a memory. Consequently, these "structures" of consciousness, i.e., perception, memory, fantasy, etc., are called *intentionalities*.

The term "intentionality" originated with the Scholastics in the medieval period and was resurrected by Brentano who in turn influenced Husserl's conception of phenomenology, who refined the term and made it the cornerstone of his theory of consciousness. The meaning of the term is complex and depends entirely on how it is conceived by a given philosopher. The term should not be confused with "intention" or the psychoanalytic conception of unconscious "motive" or "gain."

Intuition

Intuition in phenomenology refers to those cases where the intentional object is directly present to the intentionality at play; if the intention is "filled" by the direct apprehension of the object, you have an intuited object. Having a cup of coffee in front of you, for instance, seeing it, feeling it, or even imagining it - these are all filled intentions, and the object is then *intuited*. The same goes for the apprehension of mathematical formulae or a number. If you do not have the object as referred to directly, the object is not intuited, but still intended, but then *emptily*. Examples of empty intentions can be signitive intentions - intentions that only *imply* or *refer to* their objects.

Evidence

In everyday language, we use the word evidence to signify a special sort of relation between a state of affairs and a proposition: State A is evidence for the proposition "A is true." In phenomenology, however, the concept of evidence is meant to signify the "subjective achievement of truth." This is not an attempt to reduce the objective sort of evidence to subjective "opinion," but rather an attempt to describe the structure of having something present in intuition with the addition of having it present as *intelligible*: "Evidence is the successful presentation of an intelligible object, the successful presentation of something whose truth becomes manifest in the evidencing itself."

Noesis and Noema

In Husserl's phenomenology, which is quite common, this pair of terms, derived from the Greek *nous* (mind), designate respectively the real content, noesis, and the ideal content, noema, of an intentional act (an act of consciousness). The Noesis is the part of the act that gives it a particular sense or character (as in judging or perceiving something, loving or hating it, accepting or rejecting it, and so on). This is real in the sense that it is actually part of what takes place in the consciousness (or psyche) of the subject of the act. The Noesis is always correlated with a Noema; for Husserl, the full Noema is a complex ideal structure comprising at least a noematic sense and a noematic core. The correct interpretation of what Husserl meant by the Noema has long been controversial, but the noematic sense is generally understood as the ideal meaning of the act and the noematic core as the act's referent or object *as it is meant in the act*. One element of controversy is whether this noematic object is the same as the actual object of the act (assuming it exists) or is some kind of ideal object.

Empathy and Intersubjectivity

In phenomenology, empathy refers to the experience of one's own body *as* another. While we often identify others with their physical bodies, this type of phenomenology requires that we focus on the subjectivity of the other, as well as our intersubjective engagement with them. In Husserl's original account, this was done by a sort of apperception built on the experiences of your own lived-body. The lived body is your own body as experienced by yourself, *as* yourself. Your own body manifests itself to you mainly as your possibilities of acting in the world. It is what lets you reach out and grab something, for instance, but it also, and more importantly, allows for the possibility of changing your point of view. This helps you differentiate one thing from another by the experience of moving around it, seeing new aspects of it (often referred to as making the absent present and the present absent), and still retaining the notion that this is the same thing that you saw other aspects of just a moment ago (it is identical). Your body is also experienced as a duality, both as object (you can touch your own hand) and as your own subjectivity (you experience being touched).

The experience of your own body as your own subjectivity is then applied to the experience of another's body, which, through apperception, is constituted as another subjectivity. You can thus recognise the Other's intentions, emotions, etc. This experience of empathy is important in the phenomenological account of intersubjectivity. In phenomenology, intersubjectivity constitutes objectivity (i.e., what you experience as objective is experienced as being intersubjectively available - available to all other subjects. This does not imply that objectivity is reduced to subjectivity nor does it imply a relativist position, cf. for instance intersubjective verifiability).

In the experience of intersubjectivity, one also experiences oneself as being a subject among other subjects, and one experiences oneself as existing objectively *for* these Others; one experiences oneself as the noema of Others' noeses, or as a subject in another's empathic experience. As such, one experiences oneself as objectively existing subjectivity. Intersubjectivity is also a part in the constitution of one's lifeworld, especially as "homeworld."

Lifeworld

The lifeworld (German: *Lebenswelt*) is the "world" each one of us *lives* in. One could call it the "background" or "horizon" of all experience, and it is that on which each object stands out as itself (as different) and with the meaning it can only hold for us. The lifeworld is both personal and intersubjective (it is then called a "homeworld"), and, as such, it does not enclose each one of us in a solus ipse.

Husserl's Logische Untersuchungen (1900/1901)

In the first edition of the *Logical Investigations*, still under the influence of Brentano, Husserl describes his position as "descriptive psychology." Husserl analyzes the intentional structures of mental acts and how they are directed at both real and ideal objects. The first volume of the *Logical Investigations*, the *Prolegomena to Pure Logic*, begins with a devastating critique of psychologism, i.e., the attempt to subsume the *a priori* validity of the laws of logic under psychology. Husserl establishes a separate field for research in logic, philosophy, and phenomenology, independently from the empirical sciences.

Transcendental Phenomenology After the Ideen (1913)

Some years after the publication of the *Logical Investigations*, Husserl made some key elaborations that led him to the distinction between the act of consciousness (*noesis*) and the phenomena at which it is directed (the *noemata*).

- "noetic" refers to the intentional act of consciousness (believing, willing, etc.)
- "noematic" refers to the object or content (noema), which appears in the noetic acts (the believed, wanted, hated, and loved ...).

What we observe is not the object as it is in itself, but how and inasmuch it is given in the intentional acts. Knowledge of essences would only be possible by "bracketing" all assumptions about the existence of an external world and the inessential (subjective) aspects of how the object is concretely given to us. This procedure Husserl called *epoché*.

Husserl in a later period concentrated more on the ideal, essential structures of consciousness. As he wanted to exclude any hypothesis on the existence of external objects, he introduced the method of phenomenological reduction to eliminate them. What was left over was the pure transcendental ego, as opposed to the concrete empirical ego. Now Transcendental Phenomenology is the study of the essential structures that are left in pure consciousness: This amounts in practice to the study of the noemata and the relations among them. The philosopher Theodor Adorno criticised Husserl's concept of phenomenological epistemology in his metacritique *Against Epistemology*, which is anti-foundationalist in its stance.

Transcendental phenomenologists include Oskar Becker, Aron Gurwitsch, and Alfred Schütz.

Realist Phenomenology

After Husserl's publication of the *Ideen* in 1913, many phenomenologists took a critical stance towards his new theories. Especially the members of the Munich group distanced themselves from his new transcendental phenomenology and preferred the earlier realist phenomenology of the first edition of the *Logical Investigations*.

Realist phenomenologists include Adolf Reinach, Alexander Pfänder, Johannes

Daubert (de), Max Scheler, Roman Ingarden, Nicolai Hartmann, Dietrich von Hildebrand.

Existential Phenomenology

Existential phenomenology differs from transcendental phenomenology by its rejection of the transcendental ego. Merleau-Ponty objects to the ego's transcendence of the world, which for Husserl leaves the world spread out and completely transparent before the conscious. Heidegger thinks of a conscious being as always already in the world. Transcendence is maintained in existential phenomenology to the extent that the method of phenomenology must take a presuppositionless starting point - transcending claims about the world arising from, for example, natural or scientific attitudes or theories of the ontological nature of the world.

While Husserl thought of philosophy as a scientific discipline that had to be founded on a phenomenology understood as epistemology, Martin Heidegger held a radically different view. Heidegger himself states their differences this way:

> *For Husserl, the phenomenological reduction is the method of leading phenomenological vision from the natural attitude of the human being whose life is involved in the world of things and persons back to the transcendental life of consciousness and its noetic-noematic experiences, in which objects are constituted as correlates of consciousness. For us, phenomenological reduction means leading phenomenological vision back from the apprehension of a being, whatever may be the character of that apprehension, to the understanding of the Being of this being (projecting upon the way it is unconcealed).*

According to Heidegger, philosophy was not at all a scientific discipline, but more fundamental than science itself. According to him science is only one way of knowing the world with no special access to truth. Furthermore, the scientific mindset itself is built on a much more "primordial" foundation of practical, everyday knowledge. Husserl was skeptical of this approach, which he regarded as quasi-mystical, and it contributed to the divergence in their thinking.

Instead of taking phenomenology as *prima philosophia* or a foundational discipline, Heidegger took it as a metaphysical ontology: "*being is the proper and sole theme of philosophy*... this means that philosophy is not a science of beings but of being." Yet to confuse phenomenology and ontology is an obvious error. Phenomena are not the foundation or Ground of Being. Neither are they appearances, for, as Heidegger argues in *Being and Time*, an appearance is "that which shows itself in something else," while a phenomenon is "that which shows itself in itself."

While for Husserl, in the epoché, being appeared only as a correlate of consciousness, for Heidegger being is the starting point. While for Husserl we would have to abstract from all concrete determinations of our empirical ego, to be able to turn to the field of pure consciousness, Heidegger claims that "the possibilities and destinies of philosophy are bound up with man's existence, and thus with temporality and with historicality."

However, ontological being and existential being are different categories, so Heidegger's conflation of these categories is, according to Husserl's view, the root of Heidegger's error. Husserl charged Heidegger with raising the question of ontology but failing to answer it, instead switching the topic to the Dasein, the only being for whom Being is an issue. That is neither ontology nor phenomenology, according to Husserl, but merely abstract anthropology. To clarify, perhaps, by abstract anthropology, as a non-existentialist searching for essences, Husserl rejected the existentialism implicit in Heidegger's distinction between being (sein) as things in reality and Being (Dasein) as the encounter with being, as when being becomes present to us, that is, is unconcealed.

Existential phenomenologists include: Martin Heidegger (1889–1976), Hannah Arendt (1906–1975), Karl Jaspers (1883–1969), Emmanuel Levinas (1906–1995), Gabriel Marcel (1889–1973), Jean-Paul Sartre (1905–1980), Paul Ricoeur (1913–2005) and Maurice Merleau-Ponty (1908–1961).

Eastern Thought

Some researchers in phenomenology (in particular in reference to Heidegger's legacy) see possibilities of establishing dialogues with traditions of thought outside of the so-called Western philosophy, particularly with respect to East-Asian thinking, and despite perceived differences between "Eastern" and "Western". Furthermore, it has been claimed that a number of elements within phenomenology (mainly Heidegger's thought) have some resonance with Eastern philosophical ideas, particularly with Zen Buddhism and Taoism. According to Tomonobu Imamichi, the concept of *Dasein* was inspired — although Heidegger remained silent on this — by Okakura Kakuzo's concept of *das-in-der-Welt-sein* (being in the world) expressed in *The Book of Tea* to describe Zhuangzi's philosophy, which Imamichi's teacher had offered to Heidegger in 1919, after having studied with him the year before.

There are also recent signs of the reception of phenomenology (and Heidegger's thought in particular) within scholarly circles focused on studying the impetus of metaphysics in the history of ideas in Islam and Early Islamic philosophy such as in the works of the Lebanese philosopher Nader El-Bizri; perhaps this is tangentially due to the indirect influence of the tradition of the French Orientalist and phenomenologist Henri Corbin, and later accentuated through El-Bizri's dialogues with the Polish phenomenologist Anna-Teresa Tymieniecka.

In addition, the work of Jim Ruddy in the field of comparative philosophy, combined the concept of Transcendental Ego in Husserl's phenomenology with the concept of the primacy of self-consciousness in the work of Sankaracharya. In the course of this work, Ruddy uncovered a wholly new eidetic phenomenological science, which he called "convergent phenomenology." This new phenomenology takes over where Husserl left off, and deals with the constitution of relation-like, rather than merely thing-like, or "intentional" objectivity.

Technoethics

Phenomenological Approach to Technology

James Moor has argued that computers show up policy vacuums that require new thinking and the establishment of new policies. Others have argued that the resources provided by classical ethical theory such as utilitarianism, consequentialism and deontological ethics is more than enough to deal with all the ethical issues emerging from our design and use of information technology.

For the phenomenologist the 'impact view' of technology as well as the constructivist view of the technology/society relationships is valid but not adequate (Heidegger 1977, Borgmann 1985, Winograd and Flores 1987, Ihde 1990, Dreyfus 1992, 2001). They argue that these accounts of technology, and the technology/society relationship, posit technology and society as if speaking about the one does not immediately and already draw upon the other for its ongoing sense or meaning. For the phenomenologist, society and technology co-constitute each other; they are each other's ongoing condition, or possibility for being what they are. For them technology is not just the artifact. Rather, the artifact already emerges from a prior 'technological' attitude towards the world (Heidegger 1977).

Heidegger's Approach (Pre-technological Age)

For Heidegger the essence of technology is the way of being of modern humans—a way of conducting themselves towards the world—that sees the world as something to be ordered and shaped in line with projects, intentions and desires—a 'will to power' that manifests itself as a 'will to technology'. Heidegger claims that there were other times in human history, a pre-modern time, where humans did not orient themselves towards the world in a technological way—simply as resources for our purposes.

However, according to Heidegger this 'pre-technological' age (or mood) is one where humans' relation with the world and artifacts, their way of being disposed, was poetic and aesthetic rather than technological (enframing). There are many who disagree with Heidegger's account of the modern technological attitude as the 'enframing' of the world. For example, Andrew Feenberg argues that Heidegger's account of modern technology is not borne out in contemporary everyday encounters with technology.

The Hubert Dreyfus Approach (Contemporary Society)

In critiquing the artificial intelligence (AI) programme, Hubert Dreyfus (1992) argues that the way skill development has become understood in the past has been wrong. He argues, this is the model that the early artificial intelligence community uncritically adopted. In opposition to this view, he argues, with Heidegger, that what we observe when we learn a new skill in everyday practice is in fact the opposite. We most often start with explicit rules or preformulated approaches and then move to a multiplicity of particular cases, as we become an expert. His argument draws directly on Heidegger's account in Being and Time of humans as beings that are always already situated in-the-world. As humans 'in-the-world', we are already experts at going about everyday life, at dealing with the subtleties of every particular situation; that is why everyday life seems so obvious. Thus, the intricate expertise of everyday activity is forgotten and taken for granted by AI as an assumed starting point. What Dreyfus highlighted in his critique of AI was the fact that technology (AI algorithms) does not make sense by itself. It is the assumed, and forgotten, horizon of everyday practice that makes technological devices and solutions show up as meaningful. If we are to understand technology we need to 'return' to the horizon of meaning that made it show up as the artifacts we need, want and desire. We need to consider how these technologies reveal (or disclose) us.

Quietism (Philosophy)

Philosophical quietists want to release man from deep perplexity that philosophical contemplation often causes.

Quietism in philosophy is an approach to the subject that sees the role of philosophy as broadly therapeutic or remedial. Quietist philosophers believe that philosophy has no positive thesis to contribute, but rather that its value is in defusing confusions in the linguistic and conceptual frameworks of other subjects, including non-quietist philosophy. By re-formulating supposed problems in a way that makes the misguided reasoning from which they arise apparent, the quietist hopes to put an end to humanity's confusion, and help return to a state of intellectual quietude.

Quietist Philosophers

By its very nature, quietism is not a philosophical school as understood in the traditional sense of a body of doctrines, but still it can be identified both by its methodology, which focuses on language and the use of words, and by its objective, which is to show that most philosophical problems are only pseudo-problems.

Pyrrhonism represents perhaps the earliest example of an identifiably quietist position in the West. Sextus Empiricus regarded Pyrrhonism not as a nihilistic attack but rather as a form of philosophical therapy:

The causal principle of scepticism we say is the hope of becoming tranquil. Men of talent, troubled by the anomaly in things and puzzled as to which of them they should rather assent to, came to investiage what in things is true and what false, thinking that by deciding these issues they would become tranquil. The chief constitutive principle of scepticism is the claim that to every account an equal account is opposed; for it is from this, we think, that we come to hold no beliefs.

—Sextus Empiricus, Outlines of Scepticism, 12

Contemporary discussion of quietism can be traced back to Ludwig Wittgenstein, whose work greatly influenced the Ordinary Language philosophers. One of the early Ordinary Language works, Gilbert Ryle's *The Concept of Mind*, attempted to demonstrate that dualism arises from a failure to appreciate that mental vocabulary and physical vocabulary are simply different ways of describing one and the same thing, namely human behaviour. J. L. Austin's *Sense and Sensibilia* took a similar approach to the problems of skepticism and the reliability of sense perception, arguing that they arise only by misconstruing ordinary language, not because there is anything genuinely wrong with empirical evidence. Norman Malcolm, a friend of Wittgenstein's, took a quietist approach to skeptical problems in the philosophy of mind. More recently, the philosophers John McDowell and Richard Rorty have taken explicitly quietist positions.

Methodism (Philosophy)

In the study of knowledge, Methodism refers to the epistemological approach where one asks "How do we know?" before "What do we know?" The term appears in Roderick Chisholm's "The Problem of the Criterion", and in the work of his student, Ernest Sosa ("The Raft and the Pyramid: Coherence versus Foundations in the Theory of Knowledge"). Methodism is contrasted with particularism, which answers the latter question before the former.

Since the question "How do we know?" does not presuppose that we know, it is receptive to skepticism. In this way, Sosa claims, Hume no less than Descartes was an epistemological Methodist.

Experimental Philosophy

Experimental philosophy is an emerging field of philosophical inquiry that makes use of empirical data—often gathered through surveys which probe the intuitions of ordinary people—in order to inform research on philosophical questions. This use of empirical data is widely seen as opposed to a philosophical methodology that relies mainly on a priori justification, sometimes called "armchair" philosophy, by experimental philosophers. Experimental philosophy initially began by focusing on philosophical questions related to intentional action, the putative conflict between free will and determinism, and causal vs. descriptive theories of linguistic reference. However, experimental philosophy has continued to expand to new areas of research.

Disagreement about what experimental philosophy can accomplish is widespread. One claim is that the empirical data gathered by experimental philosophers can have an indirect effect on philosophical questions by allowing for a better understanding of the underlying psychological processes which lead to philosophical intuitions. Others claim that experimental philosophers are engaged in conceptual analysis, but taking advantage of the rigor of quantitative research to aid in that project. Finally, some work in experimental philosophy can be seen as undercutting the traditional methods and presuppositions of analytic philosophy. Several philosophers have offered criticisms of experimental philosophy.

History

Though, in early modern philosophy, natural philosophy was sometimes referred to as "experimental philosophy", the field associated with the current sense of the term dates its origins around 2000 when a small number of students experimented with the idea of fusing philosophy to the experimental rigor of psychology.

While the philosophical movement Experimental Philosophy began around 2000 (though perhaps the earliest example of the approach is reported by Hewson, 1994), the use of empirical methods in philosophy far predates the emergence of the recent academic field. Current experimental philosophers claim that the movement is actually a return to the methodology used by many ancient philosophers. Further, other philosophers like David Hume, René Descartes and John Locke are often held up as early models of philosophers who appealed to empirical methodology.

Areas of Research

Consciousness

The questions of what consciousness is, and what conditions are necessary for conscious thought have been the topic of a long-standing philosophical debate. Experimental philosophers have approached this question by trying to get a better grasp on how exactly people ordinarily understand consciousness. For instance, work by Joshua Knobe and Jesse Prinz (2008) suggests that people may have two different ways of understanding minds generally, and Justin Sytsma and Edouard Machery (2009) have written about the proper methodology for studying folk intuitions about consciousness. Bryce Huebner, Michael Bruno, and Hagop Sarkissian (2010) have further argued that the way Westerners understand consciousness differs systematically from the way that East Asians understand consciousness, while Adam Arico (2010) has offered some evidence for thinking that ordinary ascriptions of consciousness are sensitive to framing effects (such as the presence or absence of contextual information). Some of this work has been featured in the Online Consciousness Conference.

Other experimental philosophers have approached the topic of consciousness by trying to uncover the cognitive processes that guide everyday attributions of conscious states. Adam Arico, Brian Fiala, Rob Goldberg, and Shaun Nichols, for instance, propose a cognitive model of mental state attribution (the AGENCY model), whereby an entity's displaying certain relatively simple features (e.g., eyes, distinctive motions, interactive behavior) triggers a disposition to attribute conscious states to that entity. Additionally, Bryce Huebner has argued that ascriptions of mental states rely on two divergent strategies: one sensitive to considerations of an entity's behavior being goal-directed; the other sensitive to considerations of personhood.

Cultural Diversity

Following the work of Richard Nisbett, which showed that there were differences in a wide range of cognitive tasks between Westerners and East Asians, Jonathan Weinberg, Shaun Nichols and Stephen Stich (2001) compared epistemic intuitions of Western college students and East Asian college students. The students were presented with a number of cases, including some Gettier cases, and asked to judge whether a person in the case really knew some fact or merely believed it. They found that the East Asian subjects were more likely to judge that the subjects really knew. Later Edouard Machery, Ron Mallon, Nichols and Stich performed a similar experiment concerning intuitions about the reference of proper names, using cases from Saul Kripke's *Naming and Necessity* (1980). Again, they found significant cultural differences. Each group of authors argued that these cultural variances undermined the philosophical project of using intuitions to create theories of knowledge or reference. However, subsequent studies have consistently failed to replicate Weinberg et al.'s (2001) results for other Gettier cases Indeed, more recent studies have actually been providing evidence for the opposite hypothesis, that people from a variety of different cultures have surprisingly similar intuitions in these cases.

Determinism and Moral Responsibility

One area of philosophical inquiry has been concerned with whether or not a person can be morally responsible if their actions are entirely determined, e.g., by the laws of Newtonian physics. One side of the debate, the proponents of which are called 'incompatibilists,' argue that there is no way for people to be morally responsible for immoral acts if they could not have done otherwise. The other side of the debate argues instead that people can be morally responsible for their immoral actions even when they could not have done otherwise. People who hold this view are often referred to as 'compatibilists.' It was generally claimed that non-philosophers were naturally incompatibilist, that is they think that if you couldn't have done anything else, then you are not morally responsible for your action. Experimental philosophers have addressed this question by presenting people with hypothetical situations in which it is clear that a person's actions are completely determined. Then the person does something morally wrong, and people are asked if that person is morally responsible for what she or he did. Using this technique Nichols and Knobe (2007) found that "people's responses to questions about moral responsibility can vary dramatically depending on the way in which the question is formulated" and argue that "people tend to have compatiblist intuitions when they think about the problem in a more concrete, emotional way but that they tend to have incompatiblist intuitions when they think about the problem in a more abstract, cognitive way".

Epistemology

Recent work in experimental epistemology has tested the apparently empirical claims of various epistemological views. For example, research on epistemic contextualism has proceeded by conducting experiments in which ordinary people are presented with vignettes that involve a knowledge ascription. Participants are then asked to report on the status of that knowledge ascription. The studies address contextualism by varying the context of the knowledge ascription (for example, how important it is that the agent in the vignette has accurate knowledge). Data gathered thus far show no support for what contextualism says about ordinary use of the term "knows". Other work in experimental epistemology includes, among other things, the examination of moral valence on knowledge attributions (the so-called "epistemic side-effect effect") and judgments about so-called "know-how" as opposed to propositional knowledge.

Intentional Action

A prominent topic in experimental philosophy is intentional action. Work by Joshua Knobe has especially been influential. "The Knobe Effect", as it is often called, concerns an asymmetry in our judgments of whether an agent intentionally performed an action. Knobe (2003a) asked people to suppose that the CEO of a corporation is presented with a proposal that would, as a side effect, affect the environment. In one version of the scenario, the effect on the environment will be negative (it will "harm" it), while in another version the effect on the environment will be positive (it will "help" it). In both cases, the CEO opts to pursue the policy and the effect does occur (the environment is harmed or helped by the policy). However, the CEO only adopts the program because he wants to raise profits; he does not care about the effect that the action will have on the environment. Although all features of the scenarios are held constant—except for whether the side effect on the environment will be positive or negative—a majority of people judge that the CEO intentionally hurt the environment in the one case, but did not intentionally help it in the other.Knobe ultimately argues that the effect is a reflection of a feature of the speakers' underlying concept of intentional action: broadly moral considerations affect whether we judge that an action is performed intentionally. However, his exact views have changed in response to further research.

Predicting Philosophical Disagreement

Research suggests that some fundamental philosophical intuitions are related to stable individual differences in personality. Although there are notable limits, philosophical intuitions and disagreements can be predicted by heritable Big Five personality traits and their facets. Extraverts are much more likely to be compatibilists, particularly if they are high in "warmth." Extraverts show larger biases and different patterns of beliefs in the Knobe side effect cases. Neuroticism is related to susceptibility to manipulation-style free will arguments. Emotional Stability predicts who will attribute virtues to others. Openness to experience predicts non-objectivist moral intuitions. The link between personality and philosophical intuitions is independent of cognitive abilities, training, education, and expertise. Similar effects have also been found cross-culturally and in different languages including German and Spanish.

Because the Big Five Personality Traits are highly heritable, some have argued that many contemporary philosophical disputes are likely to persist through the generations. This may mean that some historical philosophical disputes are unlikely to be solved by purely rational, traditional philosophical methods and may require empirical data and experimental philosophy.

Criticisms

In 2006, J. David Velleman attacked experimental philosophy on the blog Left2Right, prompting a response from its defenders on Brian Leiter's blog.

Antti Kauppinen (2007) has argued that intuitions will not reflect the content of folk concepts unless they are intuitions of competent concept users who reflect in ideal circumstances and whose judgments reflect the semantics of their concepts rather than pragmatic considerations. Experimental philosophers are aware of these concerns, and acknowledge that they constitute a criticism.

Timothy Williamson (2008) has argued that we should not construe philosophical evidence as consisting of intuitions.

Other experimental philosophers have noted that experimental philosophy often fails to meet basic standards of experimental social science. A great deal of the experiments fail to include enough female participants. Analysis of experimental data is often plagued by improper use of statistics, and reliance on data mining. Holtzman argues that a number of experimental philosophers are guilty of suppressing evidence. Yet, in lumping together all people's intuitions as those of the 'folk,' critics may be ignoring basic concerns identified by standpoint feminists.

Some research in experimental philosophy is misleading because it examines averaged responses to surveys even though in almost all of the studies in experimental philosophy there have been substantial dissenting minorities. Ignoring individual differences may result in a distorted view of folk intuitions or concepts. This may lead to theoretical and strange fictions about everyday intuitions or concepts that experimental philosophy was designed to avoid akin to creating the fiction that the average human is not a man or a woman, but the average of a man and woman (e.g., the average person has one ovary and one testicle). This criticism is not unique to experimental philosophy but also applies to other sciences such as psychology and chemistry, although experimental philosophers may lack the training to recognize it.

Intercultural Philosophy

Intercultural philosophy (or sometimes world philosophy) is an approach to philosophy that had its precursors in the past but has started as a concept in the 1980s. It mostly emanates from the German-speaking parts of Europe and can be seen as a need to factor other cultures into one's own philosophical thinking and thus creating an intercultural perspective.

The Idea of Intercultural Philosophy

In the long history of philosophical thought there has always been a claim for universality although many great thinkers from the past see philosophical value only in the western tradition and oversee what other parts of the world have achieved on their own over the centuries. Eurocentric philosophers such as Georg Wilhelm Friedrich Hegel agree upon Greek being the only birthplace of philosophy and traditions such as the Chinese or Indian ones are only teachings of wisdom. Others speak of more than one birthplace and include Asian traditions.

Karl Jaspers, a German psychiatrist and philosopher, developed the theory of an axial age, referring to the period from 800 BCE to 200 BCE, during which philosophical thinking evolved in China, Indian and the Occident. Jasper's theory is widely accepted by those philosophising interculturally.

In contrast to any eurocentrism there are those philosophers who believe that there needs to be communication as well as collaboration between different traditions and cultures especially in today's global situation, given that intercultural interactions and encounters are a fact of human existence. The goal is to extend one's thinking into including other cultures, to not only consider one tradition but as many as possible such as Asian, Latin-American, Islamic, or African. It is no longer important to ask questions on your own for this would be a very regional approach. Intercultural philosophy shouldn't be an academic subject besides others but an attitude followed by everybody who philosophises. No matter what philosophical orientation, other culture's thoughts should be taken into consideration.

For Raimon Panikkar it is also important to connect religion and philosophy as they are both key elements of human reality and important to many cultures. When developing an approach to intercultural philosophy one has to abandon the idea of using only one's own ways of demonstration and description but has to include other forms such as dance, music, architecture, rituals, art, literature, myths, proverbs, folk tales and so on. A manner of meeting has to be found to allow a variety of exchange where one's own tradition can be preserved and not be forged into one big syncretism. The only way to stop cultures from being absorbed by globalisation and becoming something of a world culture, which is monoculturally predetermined, according to Fornet-Betancort, is the project of an intercultural dialogue. Others view China and Japan as an example of intercultural practise that others could learn from as they have managed to integrate Buddhism without losing their own cultural identity. Philosophers such as Wimmer and Mall postulate forms of dialogue in which all parties are on the same level ('Ebene der Gleichheit') without having any other power but the better argument.

Main Thinkers

It is not possible to name any forefathers or -mothers of intercultural thought simply because there have always been individuals in the history of philosophy that have had an intercultural approach in their theories. Although nothing defined and without much impact. Since intercultural philosophy has become a concern to more than just a couple of philosophers there are quite a few names to be mentioned. Their concept of intercultural philosophy differs according to their personal background but what they all agree upon is the practical relevance this approach presents. They each have their own suggestions for how intercultural philosophy should respond to today's situation of globalisation.

Raúl Fornet-Betancourt

Fornet-Betancourt (born 1946) is a professor in the department for missiology in

Aachen, Germany. He was born and raised in Cuba where he already came in contact with more than one culture, namely with the European or Hispanic and the African culture. His main interest lies with the Latin-American philosophy, though he has declared he doesn't research the Latin-American philosophy on his own but with help from philosophers native to this tradition. Fornet-Betancourt sees the importance of an intercultural approach in the overcoming of any eurocentrism still dominating the world. The history of philosophy shouldn't be reconstructed on the basis of the expansionary development but by means of the diversity of all cultures of humankind. But not only the past should be taken into consideration, the redesign of the present is of equal importance. Intercultural philosophy is a means for making variety heard.

Heinz Kimmerle

Kimmerle (born 1930) is professor emeritus at the Erasmus-university in Rotterdam. He intends to develop a way from colonial thinking towards a dialogue with the African philosophy based on complete equality in order to conceive of an intercultural concept of philosophy. For Kimmerle interculturality influences everything and therefore philosophy has to adapt itself to interculturality in all sub areas as to not lose its practical relevance. In his opinion philosophy of art plays an important role for it pioneers intercultural thinking.

Ram Adhar Mall

Mall (born 1937) is a professor of philosophy and teaches intercultural philosophy and hermeneutics at the university of Munich. He has systematically worked through Indian philosophy and sociology, and views himself to be an insider as well as an outsider due to his Indian heritage and Western education. For Mall interculturality derives from the overlapping of cultures that don't exist on their own. Intercultural philosophy is by no means a romantic notion for anything non-European but an attitude which has to precede philosophical thinking. Only then comparative philosophy becomes possible. Mall has worked out a hermeneutic he calls 'analogous', which moves between two hermeneutic extremes, namely radical difference and total identity. Working out overlappings despite differences enables to understand other cultures not identical to one's own. Mall pleads for abandonment of any claim to absolute right in theory as well as practice.

Franz Martin Wimmer

Wimmer (born 1942) is an associate professor at the university of Vienna. It is important to him to liberate the concept of philosophy from eurocentrism. He defines the contents of philosophy regarding to questions asked. Any tradition concerning themselves with either logic, ontology, epistemology or the justification of norms and values is indeed philosophical. Philosophy should be intercultural all the time even though it isn't yet so. It is quite a 'predicament of culturality' that philosophy claims to be universal but on the other hand will always be embedded in culture, certain means of expression and certain questions. Wimmer concerns himself with the history of philosophical thought which has to be rewritten in order to include other traditions beside the occidental. He also wants to develop ways to enable intercultural dialogues, or 'polylogues', as he calls them.

Approaches to an Intercultural Dialogue

When working with different cultures one can't just insist of one's own methods and ask everyone else to do so as well. Communications have to be adapted to this new situation. Many philosophers of intercultural thought suggest similar but nonetheless different rules or guidelines when approaching other traditions.

Polylogue

This is a concept of the Austrian philosopher Franz Martin Wimmer. He postulates that within interculturally orientated philosophy methods have to be found which disable any rash universalism or relativistic particularism. When making other voices heard, so to speak, not only should be asked what they say and why but also with what justification and due to what believes and convictions. Between radicalism and universalism there has to be a third way to carry out the program of philosophy with the help of other cultures. Wimmer calls this way polylogue, a dialogue of many. Answers to thematic questions should be worked out during such a polylogue. He drafts a 'minimal rule': never accept a philosophical thesis from an author of a single cultural tradition to be well founded. But how does a polylogue look like? Wimmer assumes an issue relevant in four traditions (A, B, C and D) for the sake of illustration. The can have one-sided influence (→) or reciprocal influence (↔). There are different models to be distinguished:

One-sided influence

A → B and A → C and A → D

In this model there is no dialogue possible. It is the goal the expansion of tradition A together with the extinction of cultures such as B, C and D. The reaction of those doesn't have to be the same. The may fiercely object or completely imitate tradition A. This is an example for eurocentrism.

One-sided and transitive influence

A → B and A → C and A → D and B → C

In this model dialogues aren't necessary as well. A continues to be the most influential culture, B ignores D, C ignores D. It may be due to the twofold influence upon C that comparative notions occur. 12

Partly reciprocal influence

There are many forms such as:

A ↔ B and A → C and A → D

or:

A ↔ B and A → C and A → D and B → C

up to:

A ↔ B and A ↔ C and A ↔ D and B ↔ C and B ↔ D and C → D

All forms here can be seen as selective acculturation. There are some dialogues or even polylogues possible, with the exception of D.

Complete reciprocal influence: the polylogue

A ↔ B and A ↔ C and A ↔ D and B ↔ C and B ↔ D and C ↔ D

For each tradition is the other quite interesting which is the consistent model for intercultural philosophy. Reciprocal influence happens based on complete equality. Of course in when it comes to practical use it might not be as carefully balanced. One tradition may be more interested in the second than the third which is a common difficulty regarding intercultural dialogue in general.

Rules of Thumb

Elmar Holenstein (born 1937) is a Swiss philosopher who concerns himself with questions regarding phenomenology and philosophy of language and culture. He observes a number of rules of thumb that make it possible to avoid intercultural misunderstandings for the most part.

Rule of logical rationality – One has to assume that thoughts not logical to oneself do not make the culture or tradition alogical or prelogical but rather that one has misunderstood them.

Rule of teleological rationality (functionality rule) – People pursue an end in what they do and don't only express themselves with logical rationality. It is easy to misunderstand if one cannot distinguish logical and teleological rationality, the literal meaning of a sentence and the goal pursued with it.

Humanity rule (naturalness rule) – Before meaningless, unnatural, non-human or immature behaviour and corresponding values are attributed to people of another culture, it is better to begin by doubting the adequacy of one's own judgement and knowledge.

Nos-quoque rule (we-do-it-too-rule) – If one encounters something in a foreign culture which one is completely unwilling to accept without contradiction, it is not unlikely that one will find comparable, if not worse occurrences in one's own culture, historical and contemporary.

Vos-quoque rule (you-do-it-too-rule) – Considering the former rule, it is no less probable that one will find persons in the foreign culture who reject the scandalous event as well.

Anti-crypto-racism rule – When people are frustrated, they are inclined to perceive their own shortcomings in magnified form in members of other groups. Crypto-racism, hidden racism, becomes manifest when one's own feeling of superiority is threatened. Foreign cultures have to be analysed to shed a revealing light on one's own culture.

Personality rule – It is possible to avoid misjudgements and tactlessness by never treating members of another culture as objects or means of research, but as research partners of equal right.

Subjectivity rule – A self-image is no more to be taken at face value than are the impressions of an outsider. According to their constitution and the kind of encounter, people tend to overestimate, super-elevate and embellish themselves, or to underestimate, diminish and denigrate themselves.

Ontology-deontology rule (›is‹ versus ›ought‹ rule) – Behaviour codes and constitutional texts do not represent conditions as they are, but as they should be according to the view of the group that have the say. Sometimes they manifest a mirror image of what is not the case but considered proper behaviour.

Depolarization rule (rule against cultural dualism) – Polarisation is an elementary means of reducing complexity and classifying things. Its primary function is not to render things as they in fact are, but rather to represent them in a manner in which they are useful. Polarisations with their simplification, exaggeration, absolutism, and exclusivity are best prevented by comparing several cultures with each other, instead of restricting the comparison to two, and by paying attention to the circumstances under which such a polar relationship between two cultures can be maintained and under what conditions it can also be detected within the cultures that are contrasted with each other.

Non-homogeneity rule – The assumption that cultures are homogeneous is a temptation to place the various eras, trends and formations to be found in them in a uni-linear order as if they are only distinguished by their degree of development and none of them has its own originality and autonomy.

Agnosticism rule – There are mysteries that will remain unanswered in all cultures and across cultures. One has to be prepared for the fact that satisfactory answers might not be found.

Gregor Paul's Basic Rules for Intercultural Philosophy

Paul is an associate professor at the university of Karlsruhe. His concerns are epistemology, logic, aesthetics and comparative philosophy as well as human rights. He has formulated 16 methodological rules regarding intercultural philosophy.

1. Ascertain similarities and make them explicit
2. Identify differences, and to describe and explain them
3. Dispel prejudices
4. Avoid mystification and exoticism
5. Assume the existence of universal, logical laws
6. To only compare equalities and to avert category mistakes
7. Avoid generalisations
8. Not to mistake parts of a tradition for the whole (e.g. identify Zen as the Eastern philosophy)

Rules regarding comparative philosophy:

1. Accept the universal validity of the common and pragmatic principle of causality as at least heuristic and pragmatic principle
2. Orient oneself on the existence of anthropological constants
3. To justify the identification of certain issues regarding to similarities and differences, in particular regarding the relevance of those identifications

Comparative philosophy should furthermore meet certain demands:

1. To explicit the underlying and guiding concept of philosophy
2. Avoid ethnocentrism and eurocentrism
3. To use terms such as 'German philosophy' and 'East' and 'West' just as abbreviation for 'philosophy formulated or developed in Germany' and 'philosophy formulated and developed in Asia'

Further common rules:

1. Multidisciplinarity and
2. Contextualisation of important examples.

Those 16 rules shall help to enable an exchange between cultures on an equal level.

Intercultural media

Beside the work of individual philosophers journals have been published to spread the intercultural thought and make as many voices heard as possible. Polylog is a journal for intercultural philosophising, published in Vienna, Austria since 1998 and offers articles mostly in the German language. Simplegadi is also a journal for intercultural philosophy, published in Padua, Italy since 1996. The journal's language is Italian. Since 2010, the Centro Interculturale Dedicato a Raimon Panikkar (Intercultural Centre Dedicated to Raimon Panikkar) publishes Cirpit Review in print or in digital format, promoting and spreading cultural events inspired by Raimon Panikkar's thought.

The blog Love of All Wisdom takes an approach similar to intercultural philosophy.

References

- Menon, Sangeetha; Anindya Sinha; B.V. Sreekantan (2014). Interdisciplinary Perspectives on Consciousness and the Self. New Youk, Dordrecht, London: Springer. p. 172. ISBN 978-81-322-1586-8. Retrieved 17 December 2015.
- Malcolm, Norman. Dreaming (Studies in Philosophical Psychology). Routledge & Kegan Paul, 1959. ISBN 0-7100-3836-4
- Husserl, Edmund. "The Crisis of European Sciences, Part IIIB § 57. The fateful separation of transcendental philosophy and psychology.". Marxists.org. Marxists.org. Retrieved 17 December 2015.
- Wheeler, Michael (12 October 2011). "Martin Heidegger – 3.1 The Turn and the Contributions to Philosophy". Stanford Encyclopedia of Philosophy. Retrieved 22 May 2013.
- Feltz, A. (2013). Pereboom and premises: Asking the right questions in the experimental philosophy of free will. Consciousness and Cognition, 22, 54-63.
- Anstey, P. & Vanzo, A. (2012). "The Origins of Early Modern Experimental Philosophy". Intellectual History Review, 22 (4): 499-518.
- Feltz, A., Perez, A., & Harris, M. (2012). Free will, causes, and decisions: Individual differences in written reports. The Journal of Consciousness Studies, 19, 166-189.

Permissions

All chapters in this book are published with permission under the Creative Commons Attribution Share Alike License or equivalent. Every chapter published in this book has been scrutinized by our experts. Their significance has been extensively debated. The topics covered herein carry significant information for a comprehensive understanding. They may even be implemented as practical applications or may be referred to as a beginning point for further studies.

We would like to thank the editorial team for lending their expertise to make the book truly unique. They have played a crucial role in the development of this book. Without their invaluable contributions this book wouldn't have been possible. They have made vital efforts to compile up to date information on the varied aspects of this subject to make this book a valuable addition to the collection of many professionals and students.

This book was conceptualized with the vision of imparting up-to-date and integrated information in this field. To ensure the same, a matchless editorial board was set up. Every individual on the board went through rigorous rounds of assessment to prove their worth. After which they invested a large part of their time researching and compiling the most relevant data for our readers.

The editorial board has been involved in producing this book since its inception. They have spent rigorous hours researching and exploring the diverse topics which have resulted in the successful publishing of this book. They have passed on their knowledge of decades through this book. To expedite this challenging task, the publisher supported the team at every step. A small team of assistant editors was also appointed to further simplify the editing procedure and attain best results for the readers.

Apart from the editorial board, the designing team has also invested a significant amount of their time in understanding the subject and creating the most relevant covers. They scrutinized every image to scout for the most suitable representation of the subject and create an appropriate cover for the book.

The publishing team has been an ardent support to the editorial, designing and production team. Their endless efforts to recruit the best for this project, has resulted in the accomplishment of this book. They are a veteran in the field of academics and their pool of knowledge is as vast as their experience in printing. Their expertise and guidance has proved useful at every step. Their uncompromising quality standards have made this book an exceptional effort. Their encouragement from time to time has been an inspiration for everyone.

The publisher and the editorial board hope that this book will prove to be a valuable piece of knowledge for students, practitioners and scholars across the globe.

Index